The Climate Bottom Line
ECONOMICS OF CLIMATE CHANGE

ISBN: 978-0-692-39353-6

The Climate Bottom Line
Economics of Climate Change

David B. King

CONTENTS

FOREWORD

My name is David B. King. I have an MA from the University of California, Santa Barbara in Economics and an MS from the University of Southern California in Systems Management. I taught a course at Santa Barbara City College in the 1970s called Environmental Economics.

The 1970s were the time of the Club of Rome, Limits to Growth, The Population Bomb and Peak Oil. In an economics course it was natural to connect these opinions about the limits to growth to the views of Parson Malthus. No matter how many historical examples of predictions of coming shortages which turned out to be false, no matter how many examples I gave of human ingenuity overcoming "finite resources" my students were firmly in the grip of a defunct economist.

So now we can see that the Population Bomb didn't go off. Many countries like Japan are in a crisis because their birthrate is at 1.4 when replacement is at 2.1. The Club of Rome has been consigned to the dust bin of history only to be replaced by Anthropogenic (man-made) Global Warning. The Malthusians, the limits to growth people, found another vehicle to slow and stop growth.

This book documents the failed, pathetic and often hilarious predictions made by climate alarmists since 1990. The Climate Models, which are the confirming experiment, have been proven wrong for 25 years. I was a programmer; a flight software engineer on the Space Shuttle Program and over my career built and used many models and simulations. This book goes into some detail about why these models are wrong, namely Garbage In, Garbage Out, or Gospel In, Gospel Out.

The book further documents the "fudge factories" of the climate scientists associated with the United Nation's International Panel on Climate Change (IPCC). They literally cooked the books, using arbitrary adjustments to produce the real man-made global warming.

I would like to dedicate this book to all the skeptics, who have been vilified and abused by climate alarmists for 26 years. I got the science, the facts and the logic from them. Any errors in this book are my own.

The curious task of economics
is to demonstrate to men how little they really know
about what they imagine they can design.

FRIEDRICH HAYEK

Chapter 1

The History

In 1988, Dr. James Hansen of NASA startled the American public with a dire prediction of anthropogenic global warning (AGW) caused by man-made emissions of CO^2. He gave three scenarios. Even in the scenario where we used no new fossil fuels after 2000, his temperature prediction turned out to be much hotter than reality. Again, in 1988, Hansen was asked by journalist Rob Reiss, how the greenhouse gases would affect the neighborhood right outside his window within 20 years (2008). "The West Side Highway (which runs along the Hudson River) will be under water. There will be tape across the windows across the street because of high winds. There will be more police cars because you know what happens to crime when the heat goes up." In 2006, Dr. Hansen also predicted a 75 ft. rise in sea level in a hundred years. The Sea Level Rate of Increase since 2004 has been at a rate of 7 inches in a century, according to a paper in *Global and Planetary Change*. Al Gore predicted a 24 ft. sea rise. The IPCC predicts 24 inches. Thus a pattern was set. A prediction of disaster which was wildly exaggerated. When called to explain the error, the true believer in climate change says, he had to choose between being effective and being honest and after all it was just a scenario of what might happen. Michael Openheimer, Princeton professor, IPCC lead author and Vice President Al Gore's science advisor, made some dramatic predictions in 1990 while working as the chief scientist for the Environmental Defense Fund. By 1995, he said, "the greenhouse gas effect would be desolating the heartlands of North America and Eurasia with horrific drought, causing crop failures and food riots". By 1996, the Platt River of Nebraska "would be dry, while a continent wide black blizzard of prairie topsoil will stop traffic on the on interstates, strip paint from houses and shut down computers". The situation would get so bad that "Mexican police will round up illegal American migrants surging into Mexico seeking work as field hands."

When confronted on his failed predictions, he refused to apologize, "On the whole I would stand by these predictions—not predictions, sorry, scenarios—as having at least in a general way actually come true," he claimed. This claim was debunked in a 2012 study in the journal *Nature* that there has been "little change in global drought in 60 years".

The Scientific Method consists of a testable hypothesis and a confirming experiment. That there would be a continent wide black blizzard of topsoil by 1996 caused by greenhouse gases is a testable hypothesis. The confirming experiment failed and thus the hypothesis failed. When your prediction fails and you say it was just a scenario, you are not making a scientific statement; you are doing what is now called Spin and what Hayek called Scientism.

The United Nation's International Panel on Climate Change (IPCC) is predicting a 3-6 degree C temperature increase in the next 100 years. These predictions of the potential of a disaster of Biblical proportions propelled climate change to the front of public policy. President Obama, Secretaries of State John Kerry and Hilary Clinton stated that it is the "biggest challenge of all that we face". A British economist recommended spending 1.2 trillion dollars a year addressing climate change. Climate change has become a totemic economic and cultural issue.

Ted Turner, the media baron who founded CNN, on PBS Charlie Rose Show on April 1, 2009, predicted that because of global warming, "We'll be eight degrees hotter in 30 to 40 years, and basically none of the crops will grow—most of the people will have died and the rest of us will be cannibals. Civilization will have broken down. The few people left will be living in a failed state—like Somalia or Sudan—and living conditions will be intolerable. Turner's solution to AGW? Keynes said that most people's politics are based on some defunct economist. In Turner's case his world view comes from Parson Malthus, who 216 years ago argued for limits on population control lest the population growth crash against limited resources. Parson Malthus was rebutted by

Ricardo in his own time, and history as rebutted Malthusian doctrine, since this limit of resources has never been reached and Malthusians have been making incorrect predictions for 216 years. Turner may not even know he is a Malthusian, and although there are many agendas behind AGW, many people who support AGW call themselves environmentalists and instead are Malthusian's wearing a green mask.

Turner went on to say on the Charlie Rose show: "We're too many people—that's why we have global warming. Too many people are using too much stuff and on a voluntary basis, everybody in the world's got to pledge to themselves that one or two children are it". China has recently announced that it is considering ending its one child per couple mandate. They have realized they are creating a demographic nightmare of an aging population with insufficient younger workers to replace them. This is the situation in Japan, the US and many other nations.

Fear of climate change was not a new idea. A *Newsweek* article, sighting a National Academy of Sciences report, warned that climate change would force economic adjustments on a worldwide scale. Worse yet, climatologists were pessimistic that political leaders will take any positive actions to compensate for climate change. This article was published in the mid-70s and was talking about global cooling.

Alarm bells had rung because the average ground temperature in the Northern Hemisphere had fallen by .5 degrees F. From 1944 to 1972. Furthermore, there had been a large increase in snow cover and between 1964 and 1972 a decrease of 1.3% in the amount of sunshine hitting the United States. *Newsweek* stated that the temperature decrease had taken us one sixth of the way to the next Ice Age. This cooling took place after WWII when CO^2 emissions were rising rapidly (IPCC 2007).

In 1971, Paul Ehrlich, a Stanford professor famous for his 1968 book, *The Population Bomb*, was another global cooling alarmist. Ehrlich said in a lecture at the British Institute of Biology: "By the year 2000 the United Kingdom will be simply a small group of impoverished islands, inhabited by 70 million hungry people. If I were a gambling man, I would take even money that England will not exist in the year 2000." Long before 2000 Ehrlich had abandoned global cooling alarmism in favor of warning that the Earth faced catastrophic global warming. Now he is warning that humans many soon be forced to resort to cannibalism. To a Malthusian, every trend is a potential crisis that allows them to warn humanity that they are trapped in a petri dish or a life boat. To them it is not about the environment, AGW or a coming Ice Age, they are haunted by a theory 216 years old which has never come true. Now they wear the green mask of Environmentalism but it is rationing limited resources like people on a life boat that they seek. Since people will not voluntarily de-develop and de-populate, it is the state that must force people to conserve. As Chairman Mao said, all the people need is two pairs of black pajamas and a bowl of rice a day.

AGW is state funded science, but it was not the left that first funded AGW to push their agenda. In 1979, Margaret Thatcher, Prime Minister of England, told the Royal Society that there was unlimited funding for studies of AGW. Thatcher had lost an election in 1974 and blamed the Coal Miners Union. She wanted to increase Britain's nuclear deterrent which was bitterly opposed by the labor party. This allowed her to kill two birds with one stone. She could support the nuclear industry and dilute the power of the coal miners by supporting AGW. She had a rather dismal reputation because she had been education minister and had cancelled the milk subsidy for school children. Thus she was called "milk snatcher" Thatcher and was not held in high regard by other international leaders.

Thatcher had a BS in Chemistry and she became a powerful voice for AGW in international Circles. Among those she influenced was Papa Bush. President Bush, the elder, also wanted the alleged support he would get for carbon free nuclear power by AGW supporters. He increased funding for Climatology from about 200 million dollars in 1992 to over two billion dollars. The head of climatology at MIT said that in 1992, Climatology went from being a quiet backwater of science, to being a hot bed of true believers chasing the funding. Predictably the flood of funding of the true believers in AGW lead to a flood of peer reviewed articles. Skeptics had their funding cut. Junior faculty learned that whatever you wanted to study from butterflies to buffalo put a little "effect of global warming" in your study and your funding got approved quickly.

The Clinton-Gore administration pushed the Kyoto Global Warming Protocol. Skeptical scientists signed a series of Declarations such as the Heidelberg, the Leipzig and the Oregon petitions (signed by 30,000 scientists) rejecting AGW. Congress refused to ratify the Kyoto Protocol and a struggle to win hearts and minds began in earnest.

Signatories to these declarations such as Dr. Fred Singer and others speaking out against AGW were put on a black list by the Society of Environmental Journalists. This society received funding from the Rockefeller foundation and other organizations pushing the AGW agenda. Their black list was titled a list of Skeptics and Contrarians. The list tries to smear the reputation by alleging ties to industry, particularly fossil fuel industries. Ironically, it has a list of true believers, trusted activists such as James Hansen, who received funding from the Heinz foundation and others.

The Leipzig Declaration, signed by many scientists such as Dr. Fred Singer, stated:

"We consider the scientific basis of the 1992 Global Climate Treaty to be flawed and its goal to be unrealistic. The policies to implement the Treaty are, as of now, based solely on unproven scientific theories, imperfect computer models, and the unsupported assumption that catastrophic global warming follows from an increase in greenhouse gases, requiring immediate action. We do not agree. We believe that the dire predictions of future warming have not been validated by the historic climate record, which appears to be dominated by natural fluctuations, showing both warming and cooling. These predictions are based on nothing more than theoretical models and cannot be relied on to construct far-reaching policies."

President Eisenhower had warned about a "military, industrial complex" in the 50s. He originally called it the "congressional, military, industrial complex". His aides convinced him take the reference to congress out. Less well known, Eisenhower warned about the increasing funding for science by government. He said that this would inevitably lead to a stifling, State Science bureaucracy.

A good example was Lysenkoism in Russia. Lysenko was Stalin's official biologist. His faulty theories were ruthlessly applied to Soviet agriculture resulting in millions of deaths from starvation caused by poor crop production. Now we have world state science (AGW), via the UN's IPCC with the funding coming from the developed world. The IPCC is a political organization whose mission is to cut CO^2 emissions, wearing the mask of science.

Scientists and the media were predicting a new Ice Age in the 70s. Freeman Dyson, Nobel Prize in Physics, said that we were 10,000 years overdue for another Ice Age. When the flood of state funding for AGW began, many began predicting catastrophic scenarios from rising temperatures 10 years later.

In March, 2000, Senior Research Scientist David Viner of the Climate Research Unit (CRU) told the UK Independent that within a few years, "snowfall would become a thing of the past in Britain". Children just aren't going to know what snow is, he said in the *Observer* article headlined "Snowfalls are now just a thing of the past".

In 2001 snowfall across the United Kingdom increased by more than 50%. In 2008, London saw its earliest snowfall since 1922. By December of 2009, London saw its heaviest snowfall in two decades. In 2010, the coldest UK winter since records began a century before covered the islands in snow.

In 2004, CRU's Viner and other experts warned that skiing in Scotland would soon become just a memory thanks to global warming. In 2013, too much snow kept many Scottish resorts closed. By 2014, the BBC reported that Scottish hills had more snow than at any point in seven decades.

The IPCC has also been hyping the snowless winter scare. In its 2001 Third Assessment Report, the IPCC claimed "milder winter temperatures will decrease heavy snowstorms". 2013 featured the fourth highest levels on record, according to the Rutgers's University Global Snow Lab. December 2014, brought a new high record for snow in the Northern Hemisphere, according to the Global Snow Lab.

THE BIG SWITCH ON SNOW

After the predictions on snow went so wrong, the CRU made the big switch. More global warming equals more snow. "If winters are warmer, it will snow more", became "settled science" overnight. The same phenomenon took place in the United States. As record cold and snow were pummeling much of North America, warming theorists contradicted all of their previous forecasts and claimed it was global warming that was causing the snow. Among them, White House Science "Czar", John Holdren. "A growing body of evidence suggests that the kind of extreme cold experience by much of the United States as we speak is a pattern we expect to see with increasing frequency". Did you follow that? As it gets warmer it will get colder. If it gets warmer, its proof of global is warming, if it gets colder, its proof of global warming. No wonder this is settled science. Now we can see why we need a Science Czar in the White House to explain this to us. Because this assertion is exactly the opposite of what the IPCC predicted in the 2001 Global Warming Report, which claimed the planet would see, "warmer winters and fewer cold spells", because of climate change. So the settled science of 2001 on snow and cold got re-settled by 2014. Do you have the feeling that if we had a string of mild winters, the settled science would get settled again?

In 2009, according to the Institute of Economic Analysis, the Climate Research Unit (CRU) cherry picked Russian Climate Data to show more warming than the 130 year meteorological stations did in their HadCRUT database. Russian met. Stations cover most of the country's territory, while the HadCRUT used data from only 25% of such stations in their calculations. The data of stations not used shows slight cooling, The IEA authors calculated that the scale of actual warming for the Russian territory in 1877-1998 was probably exaggerated by 0.64 degrees C. Since Russia accounts for 12.5% of the world's land mass, such an exaggeration for Russia alone should have an impact on the IPCC claim that the global temperature in the last century has risen by 0.76 degrees C. We will see in detail that Russia was not alone. We will learn where most of the man-made global warming came from—the "fudge factories" of climate scientists associated with the IPCC.

In January 2015, *The New York Times* headlined its article: "Last Year was the Hottest in Earth's Recorded History". Consider fig. 1 on page 9, you can see that 2014 was in the top 3% of coldest years in the last 10,000 years. In the last two million years there have been 60 ice ages. From 1400 to 1850 we had a little ice age. In the last 120,00 years we have had 20 sudden global warmings. The 1.5 degree F increase from 1850 to the present, does not even count as one of the 20, since the average decadal change in temperate is less than average at 0.16 degree C per decade, according to the Hadley Climate Research Center. We had cooling from 1900-1910, warming 1910-1940, cooling 1940 to 1980, warming form 1980 to 1998 and "pause in warming" since 1998, according to the IPCC.

The NASA climate scientist, Gavin Schmidt, director of NASA's Goddard Institute for Space Studies (GISS), who announced that 2014 was the "warmest on record" also said "greenhouse gas emissions play a major role". Several days later NASA admitted they were only 38% sure this was true. Let me repeat that in English, they were 62% sure it was not true that it was the hottest year. GISS's analysis is based on readings from more than 3,000 measuring stations worldwide and is subject to a margin of error. NASA now admits it is far from certain that 2014 set a record at all. As for the claim that "greenhouse gases played a major role" that is based on "complicated fingerprinting". Their fingerprints are that there is a lot of CO^2 in the atmosphere from humans burning coal and "signals" from climate models.

Not to be outdone by *The New York Times*, *The Associated Press* said that "nine of the ten hottest years in NOAA global records have occurred since 2000. The odds of this happening at random are about 650 million to one, according to University of South Carolina statistician John Grego. Note the, "at random" qualifier. Warm years bunch together and cold years bunch together, the AP editor is using a false assumption. Mr. Grego, the statistician with 650 million to one claim said he was instructed by the AP to assume that "all years had the same probability of being selected as one of the ten hottest years on record". This is the same as saying, if you weigh 200 pounds at some point in your life, there should be an equal chance you weigh 200 pounds at any point in your life, even when you were a baby.

Nevertheless, this AP story splashed across not only every newspaper but also every TV screen. It is now embedded in the national memory that 2014 was the hottest year on record and by implication that greenhouse gases caused it. Our national media is becoming like Pravda (The Truth) and Isvestia (The News) in cold war Russia, when the Russian people said, "there is no news in The Truth and no truth in The News."

Holocene Epoch with HadCRUT3

GISP2 Ice Core Temperature Reconstruction for Central Greeland
Data from: Alley, 2000

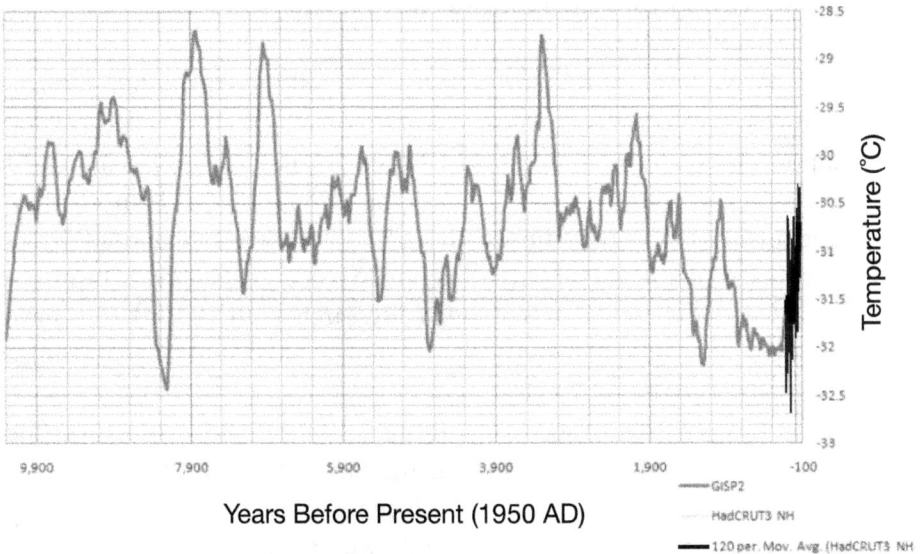

fig. 1

Chapter 2

THE SCIENCE

Is there a reason to be alarmed by the prospect of global warming? Consider that the measurement used, globally averaged temperature anomaly (GATA) is always changing. Sometimes it goes up, sometimes down, and occasionally—such as for the last 17 years—it does little that can be discerned.

According to Richard Lindzen of MIT, claims that climate change is accelerating are unsupported by facts or logic. There is general support for the assertion that GATA (global temperature) has increased about 1.5 degrees Fahrenheit since the middle of the 19th century. The quality of the data is poor, though, it is easy to nudge such data a few tenths of a degree in any direction. Several of the emails from the University of East Anglia's Climate Research Unit (CRU) that have caused such a public ruckus dealt with how to do this so as to maximize apparent changes.

The general support for warming is based not so much on the quality of the data, but rather on the fact that there was a little ice age from about the 15th to the 19th century. Thus it is not surprising that temperatures should increase as we emerge from this episode. At the same time that we were emerging from the little ice age, the industrial era began, and this was accompanied by increasing emissions of greenhouse gases such as CO^2, methane and nitrous oxide. CO^2 is the most prominent of these, and it is again generally accepted that it has increased by about 30%.

The defining characteristic of a greenhouse gas is that it is relatively transparent to visible light from the sun but can absorb portions of thermal radiation. In general, the Earth balances the incoming solar radiation by emitting thermal radiation, and the presence of greenhouse substances inhibits cooling by thermal radiation and leads to some warming.

That said, the main greenhouse substances in the atmosphere are water vapor and high clouds. Let's refer to these as major greenhouse substances to distinguish them from the anthropogenic (man-made) minor substances. Even a doubling of CO_2 would only upset the original balance between incoming and outgoing radiation by about 2%. This is essentially what is called "climate forcing".

There is general agreement on the above findings. At this point there is no basis for alarm regardless of whether any relation between the observed warming and the observed increase in minor greenhouse gases can be established. Nevertheless, the most publicized claims of the UN's IPCC deal exactly with whether any relation can be discerned. The failure of the attempts to link the two over the past 25 years bespeaks the weakness of any case for concern. Since temperatures fell from 1944 to 1972 then rose from until 1998 and have fallen since then, there is little correlation or R squared between CO_2 and temperature over this period. The EPA refuses to release to the public the studies on which they based their findings that CO_2 was harmful. Their explanation is that their studies are only for experts. The real explanation is that although correlation is not causation, they do not even have correlation and refuse to reveal this single damming fact. There is nothing so upsetting to a true believer than a beautiful hypotheses murdered by a single ugly fact.

The IPCC's Scientific Assessments generally consist of about 1,000 pages of text. The *Summary for Policymakers* is 20 pages. It is, of course, impossible to accurately summarize the 1,000 pages in just 20 pages, at the very least, nuances and caveats have to be omitted. However, even the summary is hardly ever looked at. Rather, the whole report tends to be characterized by a single iconic claim.

The main statement publicized after the 2007 IPCC Scientific Assessment two years ago was that it was likely that most of the warming since 1957 (an unusually cold year) was due to man. This claim was based on a weak argument that the current models used by the IPCC couldn't reproduce the warming from about 1978 to 1998 without some forcing, and that the only forcing that they could think of was man. Even this argument assumes that these models adequately deal with natural internal variability—that is, such naturally occurring cycles as El Niño, the Pacific Decadal Oscillation, the Atlantic Multidecadal Oscillation, etc.

Yet articles from major modeling centers acknowledged that the failure of these models to anticipate the absence of warming for the past 17 years was due to the failure of these models to account for this natural internal variability. Thus even the basis for the weak IPCC argument for anthropogenic climate change is shown to be false.

Of course, none of the articles stressed this. Rather they emphasized that according to models modified to account for the natural internal variability, warming would resume—in 2009, 2013, and 2030, respectively.

But even if the IPCC's iconic statements were correct, it still would not be cause for alarm. After all we are still talking about tenths of a degree for over 75% of the climate forcing associated with a doubling of CO_2. The potential (and only the potential) for alarm enters with the issue of climate sensitivity—which refers to the change that a doubling of CO_2 will produce in global temperature. It is generally accepted that a doubling of CO_2 will only produce a change of about two degrees Fahrenheit if all else is held constant (ceteris paribus). This is unlikely to be much to worry about. The G10 has said they will not allow global temperatures to rise more than 2 degrees C. While that does remind us of King Canute, ordering the sea not to rise, like King Canute, I think a majority of the G10 knows that they can't control the temperature of the globe but found a way to pacify the true believers by making a promise they will never have to keep.

Yet current climate models predict much higher sensitivities. They do so because in these models, the main greenhouse substances (water vapor and clouds) act to amplify anything that CO_2 does. This is referred to as positive feedback.

But as the IPCC notes, clouds continue to be a source of major uncertainty in current models. Since clouds and water vapor are intimately related, the IPCC claim that they are more confident about water vapor is quite implausible.

There is some evidence of a positive feedback effect for water vapor in cloud-free regions, but a major part of any water-vapor feedback would have to acknowledge that cloud-free areas are always changing, and this remains an unknown. At this point, few scientists would argue that the science is settled. In particular, the question remains as to whether water vapor and clouds have positive or negative feedbacks.

The notion that the earth's climate is dominated by positive feedbacks in intuitively implausible, and history of the earth's climate offers some guidance on this matter. About 2.5 billion years ago, the sun was 20%-30% less bright than now (compare this with the 2% perturbation that a doubling of CO^2 would produce) yet the evidence is that the oceans were unfrozen at the time, and that temperatures might not have been very different from today. Carl Sagan in the 1970s referred to this as the "Early Faint Sun Paradox".

For more than 30 years there have been attempts to resolve the paradox with greenhouse gases. Some have suggested CO^2, but the amount needed was thousands of times greater than present levels and incompatible with geological evidence. Methane also proved unlikely. It turns out that increased thin cirrus cloud coverage in the tropics readily resolves the paradox—but only if the clouds constitute a negative feedback. In present terms this means that they would diminish rather than enhance the impact of CO^2.

There are quite a few papers in the literature that also point to the absence of positive feedback. The implied low sensitivity in entirely compatible with the small warming that has been observed. So how do models with high sensitivity manage to simulate the currently small response to a forcing that is almost as large as a doubling of CO^2? Jeff Kiehl notes in a 2007 article from the *National Center for Atmospheric Research*; the models use another quantity that the IPCC list as poorly known (namely aerosols) to arbitrarily cancel as much greenhouse warming as needed to match the data, with each model choosing a different degree of cancellation according to the sensitivity of that model.

What does all this have to do with climate catastrophe? The answer brings us to a scandal that is considerably greater than that implied in the hacked emails from the Climate Research Unit (though perhaps not as bad as their destruction of the raw data), namely the suggestion that the very existence of warming or of the greenhouse effect is tantamount to catastrophe. This is the grossest of bait and switch scams. It is only such a scam that lends importance to the machination in the emails designed to nudge temperatures a few tenths of a degree.

The notion that complex climate "catastrophes" are simply a matter of the response of a single number—Global Temperature, to a single forcing, CO^2, represents a gigantic step backward in the science of climate. Many disasters associated with warming are simply normal occurrences which are falsely claimed to be evidence of warming. And all these examples involve phenomena that are dependent on the confluence of many factors.

Our perceptions of nature are similarly dragged back centuries so that the normal occasional occurrences of open water in summer over the North Pole, droughts floods, hurricanes, sea-level variations, etc. are all taken as omens, portending doom due to our sinful ways (as epitomized by our carbon footprint). All of these phenomena depend on the confluence of multiple factors as well.

The IPCC is a political rather than a scientific organization. Its founder stated that the purpose of the IPCC is to cut CO^2 emissions. The IPCC is not a scientific organization looking for the correct public policy. It is a political, ideological and faith based organization run by true believers who already have the answers. The IPCC states "that its only duty is to collect evidence and make plausible arguments in the hypothesis's favor". Any scientist who dissents from this orthodoxy is in danger of losing their funding. When an MIT climatologist produced a paper showing no net warming in 100 years, he lost his federal funding in a letter which said that "his findings were a threat to the future of the human race".

Michael Mann and others, supported by the IPCC, created the infamous "hockey stick". The hockey stick asserted that the Medieval Warm Period did not exist. Temperatures were low and stable until the industrial revolution. For evidence they used tree rings for proxies. Mann wrote in leaked email that he used a "trick" to "hide the decline".

His tree ring proxies showed a decline in temperatures which he hid by switching to real temperatures. He was exposed, the hockey stick was rejected, in part because we have extensive written records, paintings, etc. that the Medieval Warm Period did exist and that it was followed by a little ice age. In other words temperature wasn't low and stable and that climate has always changed. Mann still receives his federal funding and the IPCC regularly invents "plausible evidence" about Himalayan Glaciers, shrinking Antarctic Ice, hot spots in the upper atmosphere, hidden heat in the ocean, etc.

Climate change is an Orwellian Concept. Climate change was defined to be synonymous with anthropogenic global warming (AGW). In other words, every time the climate changes it is proof of AGW. One of the first techniques of propaganda is repetition. The climate changed, that is proof of climate change, over and over again.

We are constantly bombarded with the message that every weather event proves that the climate is "out of balance". This is the reason Mann and others tried to remove the Medieval Warm Period. If climate changed before the industrial age, what caused it to warm then?

Fifty international scientists have founded the Nongovernmental International Panel on Climate Change, or NIPCC. They have produced a report based on thousands of peer reviewed articles, found that "no empirical evidence exists to substantiate that 2 degree C warming produces presents a threat to planetary ecologies or environments" and that "no convincing case can be made that "a warming will be more economically costly than an equivalent cooling". Furthermore, they concluded that no unambiguous evidence exists for adverse changes to the global environment caused by human related CO_2 emissions. In particular, the cryosphere is not melting at an enhanced rate, sea level rise is not accelerating, and no systematic changes have been documented in evaporation or rainfall or in the magnitude or intensity of extreme meteorological events. These findings which are stated plainly and repeated in thousands of peer reviewed literature are not "fringe". However, many policy makers and lobby groups assert the right of the IPCC to speak for climate science.

There is correlation between CO_2 and temperature. In fact, ice core data show that temperature raises an average of 800 years before

CO^2 rises. The likely mechanism is that rising temperatures force CO^2 out of the oceans. In other words, according to ice core data, rising temperatures cause atmospheric CO^2 to rise and AGW is "standing on its head" as Hegel would say.

Even though APG is a hypothesis which is failing its confirming experiment, The Obama Administration and many state governments like California, have instituted far reaching policies to mitigate climate change. In his second inaugural address, President Obama made three assertions about climate change:

First, "Some may still doubt the overwhelming judgment of science, but none can avoid the devastating impact of raging wildfires and crippling doubt and powerful storms." A *Forbes* article in 2012 pointed out that the number of wildfires has dropped 15% since 1950, and according to the National Academy of Sciences, that trend is likely to continue for decades. Because of fire management policy of not having enough controlled burns and trying to put out every fire as quickly as possible, we have huge fuel buildups and mega fires particularly in the West. Mexico which lets many wild fires burn themselves out does not have mega fires. I once took a horseback ride with a guide in Mexico, and I was astonished that we rode right through a smoldering forest fire which no one was paying any attention to.

Second, on "droughts", a 2012 study published in the journal *Nature* said that there had been "little change in global drought over the past 60 years". The IPCC said in 2013, "droughts have become less frequent, less intense, or shorter".

Third, regarding "hurricanes and tornadoes" the President couldn't have picked worse examples. Roger Pielke, University of Colorado in a Senate hearing: "When the 2014 hurricane season starts it will have been 3,142 days since the last Category 3+ storm made landfall in the US, shattering the record for the longest stretch between US intense hurricanes since 1900. After adjusting the data for trends such as population growth and reporting, it appears that 2013 also featured the lowest number of tornadoes in the long term record".

In June, 2008, Obama said: "I am absolutely certain that generations from now, we will be able to look back and tell our children—this was the moment when the rise of the oceans began to slow and our planet began to heal". President Obama has no more power to change the climate than his experts have demonstrated to predict it.

There had been no global warming since long before President Obama took office. Global warming has been on what the alarmists call a "pause" for 18 years and counting, despite increases of CO^2 concentrations in the atmosphere. The refusal of temperatures to rise as forecasted by all of the of the UN's 73 "climate models" has discredited the models, the UN and the alleged "settled science" behind the computer forecasts. Every single model predicted more warming than has occurred, an atrocious record. Even a stopped clock is right twice a day.

The Obama administration's favorite excuse is the theory "The Ocean ate my Global Warming". In 2013, *The Associated Press*, citing leaked documents, reported that the US government had pressured the UN IPCC to incorporate that excuse, for which there is no observable evidence, into its most recent global warming report.

Chapter 3

PENTAGON CLIMATE FORECASTS

Like the UN, the Pentagon commissioned a report on climate change that also offered some highly alarming visions of the future under global warming. The 2003 document, entitled "An Abrupt Climate Change Scenario and its implications for the United States National Security" was widely cited by global warming theorists, bureaucrats and the establishment press as evidence that humanity was facing certain doom. It also served as the foundation for the claim that alleged man made climate change was actually a national security concern. By 2013, according to the report, the world should be a post-apocalyptic disaster zone. California would be flooded with Inland Seas, parts of the Netherlands "unlivable", surging temperatures, huge increases in hurricanes, tornadoes and other natural disasters were supposed to be wreaking havoc across the globe. All of that would supposedly spark resource wars and all sorts of other horrors. But none of it actually happened.

The Pentagon report even claimed there was "general agreement in the scientific community" that the extreme predictions it envisioned would come to pass, and reporters treated it as if it were a prophecy delivered to climate sinners by God Himself.

The first article about the climate report appeared in early 2004, when the report was leaked to the *UK Observer*, under the title, "Pentagon tells Bush, climate change will destroy us". In a bullet point summary at the top of the *Observer* article, journalists Mark Townsend and Paul Harris added: "Secret Report warns of rioting and nuclear war" and Britain will be "Siberian: in less than 20 years". The rest of the article was just as outlandish, going beyond what the now discredited

Pentagon report claimed. Other reporters took their cue from the *Observer* article, which in retrospect would have been hilarious piece of writing if it had not been taken so seriously at the time.

General Jack Keane, former Army Vice Chief of Staff, is quoted as calling radical Islam, "The major security challenge of our time". Yet the Commander in Chief, President Obama, in his recent 2015 State of the Union address, stated that "No challenge—no challenge—poses a greater threat to future generations than climate change".

President Obama and Susan Rice have recently emphasized again in 2015 that climate change is an important National Security issue and in a white paper gave just as many pages to climate change as they did to "Violent Extremism". President Obama, Secretary of State John Kerry and Clinton have all said climate change is the most important challenge that we face. Susan Rice said in an address to the Brookings Institute that it was "alarmist" to over react to the beheading of three Americans, ISIS in 8 countries, attacks in Paris, Boston, Ottawa, Sydney, Boko Haram and the "decimated" Al Qaeda. Can the White House really believe that a temperature increase in 50 years is the most important Foreign Policy Issue and the one we should get "alarmed" about. Isn't this just more fudge from the White House climate change fudge factory?

ARCTIC ICE

Perhaps nowhere have the alarmist's predictions been proven as wrong as at the earth's poles, the Arctic North Pole and Antarctic South Pole. In 2007, 2008, and 2009 Al Gore, the high priest of the climate cult and thus the "Goracle" publicly warned that the North Pole would be ice free in the summer by 2013 because of AGW. Speaking in Germany in 2008, he said that the entire North Polarized (sic) cap would disappear in five years".

At the UN Copenhagen Climate Summit in 2009, he said, "Some of the models suggest—that there is a 75% chance that the entire north polar ice cap, during some of the summer months could be completely ice free during the next five to seven years. We will find out".

We have found out. Contrary to the predictions of the Goracle, "some models", and his fellow alarmists, satellite data showed that the Arctic ice volume as of summer of 2013 had actually expanded by more than 50% over 2012 levels. In fact during October of 2013, sea ice levels grew at the fastest pace since records began in 1979.

Gore was not alone in making now discredited predictions about Arctic Ice. The British funded BBC, in December 2007, ran an article "Arctic Summers Ice Free by 2013." In that piece "modeling studies northern waters could be ice free in summers within just 5-8 years", some of the "experts" claimed it could happen before then, citing calculations performed by super computers that have become a standard part of climate science in recent years. Professor Wieslaw Maslowski, working with NASA, said, "Our projection of ice removal in the summer of 2013 is already too conservative".

In the real world, however, the scientific evidence the global warming theories advanced by Gore, the UN, and government funded climate scientists continue to grow, along with the ice cover at both poles. In the Arctic, data collected by Europe's Cryosat spacecraft pointed to about 9,000 cubic kilometers of ice volume at the end of the 2013 melt season; in 2012 the total volume was about 6,000 cubic kilometers.

In 2007, when Gore and others started making their predictions about imminent "ice free" Arctic summers, the average sea ice area extent after the summer melt was 4.28 million square kilometers. By 2013, on September 13, the minimum ice day for the whole year, ice levels were way above the 2007 average for the month, by an area almost the size of California. The lowest level recorded for 2013 was 5.1 million square kilometers. By late July 2014, Arctic sea ice was almost at its highest level in a decade.

Despite parroting the wild claims made five years ago, the establishment press has refused to report that Gore and his fellow alarmists were wrong. None of the scientists has lost his taxpayer funding. In fact, the establishment press is now parroting new claims that the Arctic will be ice free by 2016.

Even more embarrassing for the alarmists have been the trends in the Southern Hemisphere. The climate models and scientists predicted that rising temperatures would melt the Antarctic, by far the largest ice mass on the planet. This forecast was crucial to many of the other predictions about surging sea level and related catastrophes.

Instead, sea ice in Antarctica smashing previous record highs on a near daily basis. Sea ice area in the south is now at the highest point since records began. As of July, 2014, an area of southern ice the size of Greenland, is frozen that, based on the average, should be open waters. If both poles are taken together, there is about one million square kilometers of frozen area above and beyond the long term average.

Even the IPCC had to admit that they do not know why their climate models are so widely wrong. "There is low confidence in the scientific understanding of the observed increase in Antarctic Sea Ice extent since 1979" the IPCC admitted in the latest report.

As the *New American* reported, the desperation and denial among alarmists was illustrated perfectly in December 2014, a ship full of global warming scientists went on a mission to study how global warming was melting Antarctic ice. Instead of completing their mission, they ended up getting their vessel trapped in record-setting levels of sea ice.

Chapter 4

CO² and Global Climate Deals

In 2012 the world emitted 31 gigatons of CO_2 according to the International Energy Agency (IEA).China was first with 8 gigatons and the US was second with 5 gigatons. The US is reducing CO_2 emissions mostly because of a switch from coal to natural gas. China increased its carbon dioxide emissions 8.6% a year from 2002 to 2012. If China continues its increase at this rate for 27 years until its GDP per capita equaled the United States, it would emit 99 tons of carbon dioxide in 2041 alone, more than three times the world's current emissions. China, India and other developing nations say that the West has had its Industrial Revolution and they will have theirs. The Obama Administration is trying to reach an International "agreement" on restricting CO_2. It will be an "agreement" because they don't have the 67 votes for a Kyoto-like Treaty. In leaked emails, the Administration warned their negotiators that developing nations would demand huge payments to even sign up for unenforceable goals and promises of reductions. In 2013 China, India and Russia, the first, third and fourth largest emitter did not attend the UN meeting and along with Australia, Canada and many others clearly do not intend to trade economic growth for CO_2 reductions. Germany, while still committed to enormous expenditures on renewables is facing the highest electricity rates in the world. The only agreement that was reached was that each country would "do its own thing". China and India used to join forces to stave off pressure from developed nations to curb CO_2 emissions, arguing that high carbon dioxide levels are a problem that now-wealthy, industrialized nations created over many years and not something developing countries should be expected to solve.

China agreed for the first time in 2014 to stop increases in its emissions by around 2030. The cuts in emissions would be non-binding and thus not a treaty requiring Senate approval. The US agreed to cut emissions by 26% to 28% below 2005 by 2025. The US is cutting man made emissions because cheap natural gas is replacing coal, mostly because of market forces. In order to achieve the non-binding goal of 26% reduction, the EPA's "war on coal" would have to be continued and accelerated for 9 years after Pres. Obama's term in office.

The goal of a 26% cut in emissions is misleading, because man made emissions are 3% of total emissions. Thus a 26% cut in emissions is 26% of 3%, or a less than 1% cut. This tiny reduction would be totally offset by China alone, let alone the rest of the developing world. China only agreed to stop CO^2 increases after 2030. Prior to 2030, even cutting US emissions by 25% immediately would reduce world emissions by 4%, which would be overcome by Chinese growth in 1 to 2 years. The difference between this agreement and doing nothing is vanishingly small. The idea that if US did not make this agreement, we would be ceding the leadership on the environment to China is silly. Next time you are in Beijing or Shanghai just look out the window. If influence is based on adopting successful environmental policies, the US beats China already.

The hope of the Obama administration is that this agreement with China would put pressure on India, the third largest emitter, to agree to cuts. Instead, New Delhi is highlighting its vast differences with China, using the social, economic and emissions disparities as bargaining chip in presenting its position. India's emissions are four times lower the China's, a senior official in India's Environment Ministry said. If America is OK with the 2030 date for China, we can think about 2050, or even later.

In the last decade, China's emissions have risen sharply both in absolute and in per capita terms, reflecting rapid growth in its manufacturing economy and creating a developmental gulf with India. For example, 99.8% of China's population has access to electricity, while in India the rate is 75%, according to the World Bank.

An Indian climate policy adviser said it was Western countries that had lumped India and China together by devising a category of big emitters so that they could impose on these fast growing countries obligations that are binding. China has reached a certain stage of development. The time will come when India reaches that state too, but that time is not now, Indian officials say.

In China and elsewhere in the developing world, emissions are rising because of economic growth that is providing people with better lives. Economic development means more energy use not only for construction, transportation and manufacturing, but across the board. Even if every Chinese and Indian bought an electric vehicle, global emissions are going to rise as China, India, etc. build more cars, roads, buildings, airports etc. Persuading these nations, and indeed the US population, to incur more costs and delay improvements in their well-being in pursuit of a theory that they can control the weather by their sacrifice, is a tough sell.

In December 2014, the US, Europe and China met in Lima, Peru, to confront the gulf between their carbon-cutting ambitions and those of less industrialized nations. Officials from hundreds of countries working to negotiate a climate change wrestled with the most basic question: whether key sections of the final pact would be legally binding. The EU made it clear from the start of talks here that the climate deal set for completion in 2015 should be airtight under International Law, with legal consequences if countries stray from commitments. The EU favors an agreement with binding greenhouse gas reduction targets, said Miguel Arias Canete, the EU's energy and climate commissioner.

The US, the other giant economy active in the negotiations, is backing a hybrid approach developed by New Zealand. Under that plan, countries would be required to have an emission-cutting plan that would be enforced through domestic laws but not be binding under international law.

President Barack Obama and Chinese President Xi Jinping surprised many observers by jointly unveiling long range plans to curb CO_2 emissions. The move followed a pledge by the other big CO_2 emitting economy, the European Union, to reduce greenhouse gases as part of a new round of United Nations led climate negotiations. Officials and experts say there is little evidence most other developing economies will follow China's lead.

The risk is a revolt by the South against the North, saying that, you caused this problem and now we are going to pay the price, says Paul Bledsoe, a former US official and climate expert at the German Marshall Fund, a US based think tank.

The rift began with the 1997 Kyoto Protocol, which put the onus of emissions reductions on the highly industrialized countries that relied heavily on fossil for many decades. Since then, rapidly emerging economies with big populations and expanding middles classes have overtaken many developed nations in burning coal and other fossil fuels.

Many of these countries, including a number of energy producers, don't want to imperil their growth if richer countries will make the cuts needed to keep global temperatures from rising more than 2 degrees Celsius, a goal world leaders set at the 2009 Copenhagen Summit. This attitude is aggravated by the spectacle of Secretary of State John "Five Homes" Kerry jetting into Indonesia on a column of CO^2, to lecture the Indonesians on reducing their sinful carbon footprint. Secretary of State Kerry earned his fortune, his five homes and the carbon footprint of a small third world country, the old fashioned way, by marrying the heiress to the Heinz fortune.

American and other officials are seeking to blur the old, two-way standoff by encouraging every nation to develop its own emissions reduction plan to fit its own growth expectations and abilities. The Obama administration is clearly offering the China deal. Pick a date in the future, 2030 for China, India wants 2050, say you will cut emissions then, the US will cut emissions now and we have already pledged billions to be paid now, if you will pledge to cut later. Since none of these goals and pledges is enforceable, they allow the "messaging" that progress is being made.

Dr. Benny Presser, the director of the Global Warming Policy Forum in December 2014 welcomed the non-binding and toothless UN climate agreement which was adopted in Lima.

"The Lima agreement is another acknowledgement of international reality. The deal is further proof, if any was needed that the developing world will not agree to any legally binding caps, never mind reductions of their CO^2 emissions. As seasoned observers predicted, the Lima deal is based on a voluntary basis which allows nations to set their own voluntary CO^2 targets and policies without any legally binding caps or international oversight. In contrast to the Kyoto Protocol, the Lima deal opens the way for a new climate agreement in 2015 which will remove legal obligations for governments to cap or reduce CO^2 emissions. A voluntary agreement would also remove the mad rush into unrealistic decarburization policies that are both economically and politically unsustainable". Lord Nigel Lawson, Chairman of the Global Warming Policy Forum, added: "The UD's unilateral Climate Change Act is forcing British industry and British households to suffer an excessively high cost of electricity to no purpose. Following Lima, it is clearer than ever the Act should be suspended until such time as a binding global agreement has been secured".

There is a growing consensus even among AGW zealots, that there is no hope of legally binding or enforceable deals. After the high-level phase of the Lima talks got off to a slow start, Sec. of State Kerry made a surprise visit. Urging countries not to get bogged down in "debates over who should do what". "I am confident we can we can rise above the debates that have dragged us down."

There was more happy talk from Tim Groser, New Zealand's trade and climate minister. "The number one requirement is participants, not depth of cuts. We're not going to get rid of oil in the next 50 years, for God's sake." Enforceable deal or not, there will be no shortage of government officials who want an all-expense paid trip to Paris in 2015. The Kyoto Protocol, if it had been carried out by every nation, would have cut temperatures in 100 years by a fraction of a degree. It is obvious that when the Paris Climate Deal is announced, if you do the CO^2 math, it will promise to do even less.

Diplomacy is easy as long as the appearance of progress matters more than results and a bad deal. Such a world view appears to be the price of victory at the roaming international climate talks which has such a deal to take to Paris in 2015. The 196 nations in Lima followed the familiar pattern of previous United Nations installments Durban, Cancun and Copenhagen" The event nearly collapsed amid the irreconcilable demands of rich and poor countries, only for the negotiators to agree in overtime to negotiate more at a later date. The greens are no less optimistic about planetary carbon salvation, or at least an all-expense paid two weeks in Paris.

Supposedly the Lima deal is a breakthrough because the developing nations that were exempted in the 19 previous rounds will promise to reduce emissions for the first time. Less developed countries, those outside the OECD are responsible for 57.5% of global emissions over the last five years. Without their participation, atmospheric CO^2 will continue to accumulate whatever the West does.

The problem for the climate lobby is that the non-OECD carbon share continues to grow because fossil fuels are helping to lift a generation out of poverty. The middle class is growing in China, while some 400 million people in India and 550 million in Africa still lack electricity. Such basic necessities have been a higher priority than a threat that controversial climate models predict will arrive decades from now.

So the irony is that the climateers in Lima allowed the developing world to volunteer meaningless carbon promises. Under the four page Lima agreement and its 39 page annex, all countries will receive a United Nations invitation to define a carbon reduction program of their own choosing, whenever they're ready, with no specific goals or consequences if they don't comply.

Chinese and Indian delegates demanded that every use of the word "shall" be changed to "may", or else they would walk. They even succeeded in stripping language that countries should commit to providing verifiable, transparent, consistent and complete, accurate and comparable information.

In other words, the countries, India, China, etc. who insisted on these changes, now have the UN's permission to do whatever they want and call it fighting climate change. Meanwhile, demanding the billions of dollars in aid, they have been promised. The predictable result will be anti-fossil fuel masochism in the US and Europe in return for vague, unenforceable promises.

This agreement was supposed to be different because the US and China agreed to limit emissions. However, Lima has exposed the reality of that deal: President Obama said he will impose such limits on Americans without a vote in congress, while the dictatorship that supplies 27% of annual global carbon output made a no-detail pledge that its emissions would maybe stop growing after 2030. (The US contributes 17 %.)

The Lima Climate Conference has opened the one-sided bargain to all comers, bringing the world no closer to an anti-carbon policy. So ordinary Americans and Europeans will be forced to accept lower economic growth today and higher energy prices, and these sacrifices will make no difference to the problem they are meant to solve.

Chapter 5

METHANE AND BIO FUELS, HOW RENEWABLES HARM THE ENVIRONMENT

The White House has announced plans to impose a new set of regulations on the oil and gas industry, targeting the emissions of methane. The basis is shoddy climate science, as propagated in various UN-IPCC reports. These claim that global warming potential of a methane molecule is about 50 times that of CO^2 and that climate forcing from growth of atmospheric methane is about 20% of carbon dioxide's. As Dr. Fred Singer writes:" They (the IPCC) made two scientific errors, as can be readily shown. They ignored the fact that the infrared absorption bands of atmospheric water vapor cover those of methane, one cannot absorb the same radiation twice. Further, the methane bands are located far from the peak of the surface heat emission spectrum, where there is little energy available to be absorbed. I don't know how the IPCC got its numbers—but they are wrong".

Singer published early estimates of anthropogenic (man-made) production of methane and its contribution to stratospheric water vapor (*Nature*, 1971) since confirmed by data.

The US has pledged to reduce greenhouse gases through regulation by the EPA, this includes a war on cow farts. The EPA blames bovine burps and flatulence for global warming. According to the EPA all of US agriculture accounts for just 8% of our greenhouse gas emissions, with by far the largest share coming from soil management—that is, crop farming. (Please note: 8% of 3% of total emissions) A Union of Concerned Scientist report concluded that about 2% of the US greenhouse gases can be linked to cattle and that good management would diminish it further. The primary concern is methane, a potent greenhouse gas. But methane from cattle, now under vigorous study and regulation writing, can be mitigated in several ways. Australian research shows that certain supplements can cut methane from cattle by half. Will the EPA mandate Beano for cattle? Things as intuitive as good pasture management and as obscure as robust dung beetle populations have all been shown to reduce methane. Perhaps we will have an EPA regulation, a dung beetle in every cow patty.

At the same time, cattle are key to restoring carbon to the soil. One tenth of all human caused carbon emissions since 1850 have come from soil, according to Richard Houghton of the Woods Hole Research Center. This is due to tillage, which releases carbon and strips the Earth of protective vegetation, and to farming practices that fail to return nutrients and organic matter to the Earth. Plant covered land that is never plowed is ideal for recapturing carbon through photosynthesis and for holding it in stable forms.

Most of the world's beef cattle are raised on grass. Their pruning mouths stimulate vegetative growth as their trampling hoofs and digestive tracts foster seed germination and nutrient recycling. These beneficial disturbances, like those once caused by wild grazing herds, prevent the encroachment of woody shrubs and are necessary for the functioning for grassland ecosystems.

Research by the Soil Association in the UK show that if cattle are raised primarily on grass and if good farming practices are followed, enough carbon could be sequestered to offset the emissions of all UK beef cattle and half its dairy herd. Similarly, in the US the Union of Concerned Scientists estimates that as much as 2% of all greenhouse gases (slightly less than what is attributed to cattle) could be eliminated by sequestering carbon in the soils of grazing operations.

These studies, unfortunately, have not and will not deter federal and state regulators from going after cattle operations. There is a strong vegan wing on the environmental coalition who wants all cattle raised for beef to be eliminated. They are untroubled by the fact that wild animals like deer, elk, bears etc. all produce methane. In the faith based wing of the Climate Change Coalition, natural emissions of methane, oil seeps, CO^2 from vegetation etc., even though they constitute 97% of total emissions are natural and therefore not "pollution". Ten thousand gallons of oil seep off the coast of the University of California, Santa Barbara every day. This oil is chemically identical to oil pumped from platforms. It sometimes forms a 50 mile sheen that can occasionally be seen from planes taking off from the Santa Barbara Airport. This seep oil gets on the beach and oils birds. But many of the academics have said publicly that since it is natural it is not pollution. This is opinion, not facts, logic or science, but it is announced as if it is the one true faith. Gaia Akbar! (Earth is great!)

Grass is also one of the best ways to generate and safeguard soil and to protect water. Grass blades shield soil from erosive wind and water, while its roots form a mat that holds soil and water in place. Soil experts have found that erosion rates from conventionally tilled agricultural fields average one to two orders of magnitude greater than erosion under native vegetation, such as what's typically found on well managed grazing lands.

Nor are cattle voracious consumers of water. Some environmental critics of cattle assert that 2,500 gallons of water are required for every pound of beef. But this figure (or the even higher ones often cited by advocates of veganism are based on the most water intensive situations. Research at the University of California, Davis shows that producing a typical pound of beef takes about 441 gallons of water per pound— only slightly more water for a pound of rice and beef is far more nutritious. Eight pounds of broccoli for the B vitamins in one ounce of beef, for example.

Eating beef also stands accused of aggravating world hunger. This is ironic since a billion of the world's poorest people depend on livestock. Most of the world's cattle live on land that cannot be used for crop cultivation, and in the US, 85% of the land grazed by cattle cannot be farmed, according to the US Beef Board.

The bovine's most striking attribute is that it can live on a simple diet of grass, which it forages for itself. And for protecting land, water, soil and climate there is nothing better than dense grass. As we consider the long term prospects for feeding the human race, cattle will rightly remain an essential element.

A surge of new federal regulation is headed for the oil and gas industry in 2015 with a crackdown on methane. The administration is targeting offshore drilling and oil trains but the methane rule is likely to be among the worst. The non-crisis that it will purport to solve is already well in hand thanks to industry innovation. The Environmental Protection Agency is going ahead anyway.

Methane is a more potent greenhouse gas than carbon dioxide, though CH4 is far less prevalent than CO^2, and has a much shorter atmospheric life. The real reason the methane has become an obsession with the Malthusians wearing the green mask is that it sometimes leaks when extracting or transporting oil and gas. Oil and gas are the key ingredients that are causing a cancerous growth which is consuming the finite resources in their Petri dish, and which must be stopped. Methane can be a pretext for interfering with and raising the costs of drilling.

But this means willfully ignoring the plunge in US methane. Overall emissions fell 4.7% between 1990 and 2008 and 6.3% between 2008 and 2012. The most recent year an estimate is available in the EPA's greenhouse gas inventory. Natural gas is the source of less than a third of the total, the next largest being "enteric fermentation" or livestock flatulence. These emissions (cow farts) rose 2.3% over 1990-1912.

Methane reduction in the drilling industry has been dramatic. Methane emissions from natural gas systems fell 14.3% from 2008 -2012. Since 2011 the EPA has also specifically measured leaks from hydraulically fractured natural gas wells. By 2013 those dropped 73% more than any other industrial sources.

These are the same years when the US became the world's natural gas leader, with production increasing by nearly fourfold since 2008. The US added 600,000 miles of gas pipeline, a 30% increase, utilities substituted gas for coal on a massive scale, which caused US CO^2 emissions to fall, and the economy grew. Methane emissions nonetheless fell. The fact that the EPA is still going after methane reveals that the goal is not methane; it is to stop growth and de-develop industry. The US will never transition to expensive renewables while we have cheap, abundant fossil fuels.

In December 2014, engineers at the University of Texas—funded in part by the Environmental Defense Fund, which is pushing federal methane regulation—published the most exhaustive study to date of methane emissions and fracking. The UT team found that the leakage rate as measured in the field was not only lower the EPA's assumptions but had also fallen 10% year over year.

Methane is not a byproduct of burning natural gas like CO^2. The hydrocarbon mixture laymen call natural gas is primarily composed of methane itself, and leakages means drillers and transporters are losing the valuable commodity they are trying to sell. The economic incentive to capture CH4 has translated into rapid technological progress, and emissions are declining in every significant basin from Texas to North Dakota to Appalachia as more efficient techniques spread throughout the industry.

The EPA hasn't released details about its looming methane rule, but at best it will be expensive and redundant. The Malthusians with the green mask are demanding that the agency mandate CH4 reductions and impose technology performance standards like pneumatic controller to throttle back production. This will help achieve the real goal, raise the price and lower the supply of natural gas in order further the transition to renewables.

BIOFUELS

Biofuels sound good, "fuel that grows itself" and "a great use for wood waste", in reality, they are uneconomic and harm the environment. The first problem is that there is a limited supply of waste biomass and power plants need a steady supply of fuel. So power producer's need to rely on virgin feedstock. In an article in the journal *Economist*, Poland and Finland wood meets 80% of renewable demand. In Germany, wood makes up to 38% of non-fossil fuel power consumption. Where does this wood come from? A US pellet producer, Enviva is sourcing its wood from the clear cutting of bottomland hardwood forests—some of the most biodiverse temperate forests and freshwater ecosystems in the world. Canada exports about 1.3 million tons of wood pellets made from forests.

As for being carbon neutral, boreal forests grow slowly and model simulations from the journal, Climate Change indicate that harvest of a boreal forest will create a "biofuel carbon debt" that takes 190 to 340 years to repay. Pellets made from forests are carbon neutral after a couple of centuries. In order to provide electric power in Europe from "renewables" in order to "fight climate change" US and Canadian forests are being cut down, often at an unsustainable rate. This results in the destruction of valuable habitat and loss of ecosystem diversity. The power being used in Europe by Greenpeace and others to fight the tar sand's alleged destruction of boreal forests is provided by the cutting down and grinding up of actual Canadian boreal forests.

THE BIOFUEL ETHANOL

According to *Forbes*, in 2000 over 90% of the US corn crop went to feed people and livestock, with less than 5% used to produce ethanol. In 2013, in 2013, 40% went to produce ethanol. Food for 500 million people was pulled out of the food chain to run vehicles. *Forbes* also says that Brazil is clear-cutting a million acres of tropical forest a year to produce a biofuel and shipping most of it too Europe to meet renewable targets. The net effect is about 50% more carbon emitted by using these biofuels than using petroleum fuels. A recent report in Science says that corn based ethanol nearly doubles greenhouse gas emissions over 30 years and increases greenhouse gases for 167 years. Thus it will be carbon neutral in 167 years. Switch grass increases emissions by 50%. Another article in *Science,* indicates that converting rainforests, peat lands, savannas to produce food crop based biofuels in Brazil, Southeast Asia and the United States creates a "biofuel carbon debt" by releasing 17 to 420 times more CO_2 than the annual greenhouse gas reductions that these biofuels would provide by displacing fossil fuels.

THE BIOFUEL PALM OIL

An article in *Ensia* reports that in 1985, Indonesia had less than

2,500 square miles of palm oil plantation, 20 years later, they covered 21,621 square miles, and by 2025 the Indonesian government projects plantations will cover at least 100,000 sq. miles. Since there are over 1,000 palm oil plantations in Indonesia, this is the equivalent of 220,000 cars a year, this is in addition to the palm oil plantation's biofuel carbon debt of over a century.

Biofuels, when they are used in the manner they have been used to date, are destroying our shared ecological inheritance. Each year thousands of acres of forests in South and Central America and Southeast Asia are being clear cut or burned to in order to free up space for the production of these supposedly "carbon neutral" fuels. Yet, these fuels can only be considered carbon neutral if you look at them over hundreds of years of carbon debt. Ecosystems need resiliency and the destruction of habitat reduces resiliency and increases the risk of ecological collapse in degraded ecosystems. Moving the calories used in biofuels out of the human food chain increases costs for food. Well meaning, but fact, logic and science blind activists for AGW and biofuels need to learn what there slogans are actually accomplishing.

Chapter 6

COAL

The push for a global climate deal is leading to calls for large cutbacks in the use of coal, provoking stiff pushbacks from coal producers and exporting countries. Many developing countries, which are cutting coal's share of electricity production at home and promoting other fuels, are discouraging governments and international institutions from investing in new coal plants that would feed the growing energy appetitive of poorer countries.

"Coal is a source of energy that the past was based on, not a source that the future will be based on", said Susan McDade, deputy assistant administrator of the United Nations Development Program, which finances projects in developing countries.

The head of the Chinese delegation noted that Tokyo wasn't following its example of submitting an early target for reduced carbon dioxide emissions, in effect highlighting Japan's reliance on coal after the Fukushima nuclear disaster. Japan's envoy said the coal dependence situation is going to be improved.

But the anticoal rhetoric is putting coal's champions on the defensive and stirring political opposition, as industry giants seek to promote coal as a cheap fuel for developing countries.

Australian Foreign Minister Bishop defended her government's explicitly pro-coal stance, saying the county has committed more than $300 million dollars to developing low-emissions coal technology.

But after leading industrialized countries—including China, the biggest source of CO^2 emissions—pledged to make cuts, some Lima participants worry that developing countries, including India will pick up the slack, potentially boosting coal consumption in inefficient plants with high levels of pollution and CO_2 emissions.

To keep richer countries extra coal from being burned in poorer countries, the US, the World Bank, United Nations and others are curtailing funding for coal fired electricity. In the US, Republicans used legislation that would fund the Gov. in 2015 to hinder President Obama's efforts to limit investments by US agencies in overseas coal fired power plants. They are also seeking to use gains in congressional elections to roll back the Environmental Protection Agency's power plant regulation. There currently exists no carbon capture technology that is even remotely economic. Thus the requirement for carbon capture is tantamount to an eventual plant closure. Whether or not the EPA can impose a technology that doesn't exist and thus for which they are no cost estimates will probably be decided by the Supreme Court in 2015.

The US has doubled coal exports in the last decade, and miners around the world want to preserve their ability to extract and export coal, even if political pressure makes it harder to use the fuel at home. Overall, new government policies are expected to help slow the growth in coal demand to 0.5% per year through 2040, compared with growth of 2.5% during the past 30 years, according to the International Energy Agency.

A global pact on climate change intending to cut coal demand and production faces significant challenge for coal producers and exporters, like coal states such as Kentucky. The top countries for coal as a percentage of electricity production in 2012: Mongolia (95%), South Africa (93%), Poland (83%), China (81%), India (71%), Australia (69%), Israel (61%), Indonesia (48%), Germany (44%), US (38%), UK (39%), Japan (21%). Special mention should be paid to Germany. Under Chancellor Merkel the Germans have the most aggressive CO_2 reduction goals in the world. However, by investing in a massive offshore wind farm in the North, without obtaining the right of way for transmission lines to the industrial south, Merkel's energy revolution is over budget by billions, behind schedule by years, and by banning nuclear has given Germany the highest priced electricity for manufacturing in the world while actually increasing CO_2 emissions by increasing coal use.

The top coal exporters in 2013 in millions of metric tons, Indonesia (426), Australia (336), Russia (143), US (107), Columbia (24), South Africa (72), Canada (37). Each of these nations is unlikely to sign a global treaty on climate change which only asked for unenforceable goals, let alone one with real sanctions. Although the EU is asking for enforceable limits on CO^2 emissions, what sovereign nation would submit to fines etc. from the World Court, the United Nations or some other global state entity? The US has pledged billions of dollars to aid other countries in CO^2 emission reductions. It is unlikely that a single dollar will be actually be paid.

US emissions have decreased in recent years due to the increasing use of natural gas. We haven't done any of the hard parts of reducing emissions—steps that will eliminate jobs and hurt growth. Cutting energy use is technically difficult and expensive. Carbon fuels are the source of more than 80% of US energy and everything that we make and consume has energy embedded in it. The vast majority of our buildings cannot be significantly more energy efficient just by adding insulation and caulking the windows. Expensive modifications will be required.

At the same time, shifting to renewables or dramatically increasing mandates for efficiency will make everything cost more. Just shifting our transportation fleet to biofuels, hydrogen, electric, etc. would require extensive expansion of infrastructure, all of which would require huge public expenditures. Since fossil fuels, gas, diesel and natural gas would still be the most cost effective transportation fuels, even huge expenditures and subsidies on substitutes would still leave us with a transportation fleet running on fossil fuels.

The US cannot afford such costs right now. Overtime, however, technological progress will cut the price of reducing greenhouse gases. Yet another reason to take the time needed to forge a global agreement. On the other hand, if it continues to cool, by waiting we will have more growth which will give us the resources to deal with climate change, hot or cold.

CO^2

Global temperatures have varied by plus 6 and minus 4 degrees from today's temperature. CO^2 was 15 times today's level 550 million years ago. Our food crops, like wheat, evolved with much higher levels of CO^2. Greenhouses add CO^2 up to levels 3 times current levels, because todays plants are CO^2 starved. We exist because our evolutionary forbears adapted and survived. Strong advocates of AGW reject adaption because they seem to believe that even considering alternatives to cutting CO^2 will weaken the publics will to endure the pain of pretending to cut global CO^2 emissions. AGW advocates and governments talk about reducing greenhouse gas, but what they mean is CO^2. Few know that CO^2 is less than 4% of all the greenhouse gases and the human portion of CO^2 emission is 3% of total CO^2 emissions. The human contribution is within the error range of two of the natural sources, Oceans and Ground Bacteria. In other words, if every human being left the planet except one scientist remained to measure the difference in atmospheric CO^2 they would not be able to measure any difference.

Chapter 7

ADAPTION

It's time to change the Climate Change Agenda. It's time to adapt to reality. In the 70s the correct public policy was to ignore the predictions of a coming Ice Age. Now the correct public policy is to ignore the predicted catastrophes in order to maximize economic growth. Then we will have the resources to adapt to whatever the future brings.

In 2013 climate related deaths were at a record low, while AGW alarmists were saying were saying it was the worst climate in history. In 1931, bad weather killed 3 million people. Industry and the fossil fuels that drive it are lowering climate deaths year after year. Cold weather has always killed far more people than warmer weather. For every degree that is warms, net deaths will fall. For every degree that temperature cools, net deaths will rise. Humanity owes its current ability to survive harsh winters, arid deserts and other naturally dangerous environments to the same fuels AGW activists now condemn. We have the luxury of being able to absorb a certain amount of climate related damage so we can live in many varied climates, warm and cool. Environmentalists seem to have the notion that without human interference, the planet is perfect. If you went to someone 200 years ago and asked them, do you have a perfect climate? They would think you were crazy. They were terrified of climate, because climate often doesn't give you the resources that you need. It doesn't give you water when you need it. This was the time of the little ice age. If you ran out of wood, you froze to death.

The US does a national assessment of climate change impacts. It lists huge costs from droughts, floods, storms, disease etc. These studies always conclude that the last two decades have the worst climate in recorded history. Why then has global life expectancy for men and women increased by about six years, according to one of the most comprehensive studies done so far in the Journal Lancet. Many AGW studies claimed an epidemic of malaria because of the climactic impact of global warming. Malaria deaths have fallen, in fact death from infectious diseases as a whole has dropped by about 25%. The IPCC for the last twenty years has through their acolytes in the media dominated climate impact information by constantly inventing a catastrophe that might happen and then claiming it has already happened. The first tool of propaganda is endless repetition. In a process that has come to be called a climate audit, the truth will out. There have already been climate trials in England and in New Zealand. The Film "Inconvenient Truth" was found to be propaganda in the High Court in London. In New Zealand, scientists were caught cooking the temperature books by a process they called homogenization, which turned cooling raw data into warming adjusted data. There are over 90 lawsuits in the US, trying to get the raw data and adjustments the EPA uses to justify its war on coal agenda. The EPA is refusing to release the data because they say it is only for their experts. It is only a matter of time before this data and how it has been massaged becomes public. Then there will be climate trials in the US and it will not be skeptics in the dock.

Instead of talking about climate change of which there will always be some, we should focus on climate catastrophe, which should be defined as climate that actually kills people. Those catastrophes, measured in lost lives, are getting rarer. "The dogma that man is ruining the planet rather than improving it is a religion, a source of prestige and political power and a career for far too many people" wrote Alex Epstein.

Most of the changes humans make to our environment are desirable changes that help us live longer and more comfortably. Consider the hated plastic water bottle. Of course we should eliminate and clean up all plastic trash. Already there are substitute plastics that are made from plant material, which naturally biodegrade. But plastic water bottles were a major advance in public health in many places like Burma and Peru, the first source of known to be clean water for millions of people.

If we regard Nature as pristine and think it must never be altered, and take draconian steps like banning fossil fuels, we will die young and lead miserable, difficult lives. Think of industry as something that is mostly good for us, with a few side effects that aren't. Fossil fuels are a little like antibiotics. It's good to draw attention to minor side effects, but it would be crazy to abandon all treatment because of them. Fossil fuels are no catastrophe. They contribute to health, a longer life and economic growth.

Chapter 8

DROUGHTS

The crucial, unsettled scientific question for policy is, "How will the climate change over the next 30 to 100 years under both natural and human influences". Answers to that question at the global and regional levels should guide our choices about energy and infrastructure. In California, the drought cycle is heavily influenced by ocean temperature. Warm sea water is called El Niño and cooler water is called La Niña. California is currently in a severe drought supported by a delay in the normal arrival of warm water. Media propaganda usually blames drought on climate change contrary to the IPCC which predominately associates it with increased precipitation.

The current science cannot explain why the warm ocean water was delayed, when the El Niño did arrive it was weak (not very warm ocean temperature) and thus there is no public policy guidance. This, in general, is the case; the science is so immature there are no public policy implications. A report issued in December of 2014 by the National Oceanic and Atmospheric Administration said natural variations—mostly a La Niña (cold water) weather oscillation—were the primary drivers behind the drought that had lasted for three years. Study lead author Richard Seager of Columbia University, said the paper has not yet been published in a peer-reviewed scientific journal. He and NOAA's Martin Hoerling said 160 runs of computer models show heat-trapping gases should slightly increase winter rain in parts of California, not decrease. The conditions of the last three winters are not the conditions that climate change models say would happen.

He said the La Niña (cold water), which is the flip side of the warming of the Central Pacific Ocean (El Niño), can only be blamed for about one-third of the drought. The rest of the causes can be from just random variation.

Kevin Trenberth, head of climate analysis at the National Center for Atmospheric Research, said this report completely fails to consider what climate change is doing to water in California. He said the work completely misses how hotter air increases drying by evaporating more of it from the ground. In droughts, extra heat from global warming enhances the drying in a feedback effect, Mr. Trenberth said. {The IPCC admits it does not have good data for soil moisture, which is a critical requirement for climate models. The US is launching a satellite which for the first time will acquire global data for soil moisture. Absent this critical data item and many more which we will document, current climate models cannot work.}

But Mr. Hoerling said that is less of a factor in California because it is so near the ocean and its rain comes in storms coming off the Pacific. Peer reviewed studies are divided on whether the drought can be blamed on climate change. Others published this year point more directly to changes in pressure of the Pacific that blocked rain from coming into California, but Hoerling and Seager dismissed them as not adequate.

Mr. Hoerling, of NOAA, who specializes in the complicated field of studying the cause of climate extremes, in the past, has downplayed other scientists' claims that regional droughts are caused by man-made warming. However, Mr. Hoerling acknowledges that climate change is happening and has produced past studies attributing strange weather— such as more frequent Mediterranean droughts—to heat trapping gases from the burning of fossil fuels.

We will show later in the section on climate models, that the IPCC admits that the models do not contain accurate data on moisture. Thus climate models that are being used to study droughts etc. do not contain the variable required to confirm or deny not only droughts but climate period. Climate models are crippled by garbage data in, garbage data out, or in the case of climate change models, gospel in, gospel out.

Scientists can't even agree on how bad the drought is. Mr. Hoerling said the drought isn't even in the top five worst for California. But a new peer-reviewed study in the journal Geophysical Research Letters by researchers at the University of Minnesota and the Woods Hole Oceanographic calls this "the most severe drought in the last 1,200 years. This study was based on tree ring data, which has a nasty reputation thanks to Michael "hockey stick" Mann.

Deke Arndt, climate monitoring chief for NOAA's National Climatic Data Center, said by some drought measures, the current California drought is slightly more intense than, but still comparable to, the late 1970's episode. "I'd put them at 1A and 1B on the list of historical multi-year drought episodes affecting California in modern times."

Please note the disagreement on climate data. Why do they have to run the climate models 160 times and then average the results? Do you run TurboTax 160 times and average the results? What public policy is indicated by this discussion on the California drought? Shall we continue to impose draconian costs in increased energy prices etc. based on an immature science in order to pretend we are controlling climate change?

Al Gore on coral and AFW: "Coral reefs all over the world, because of global warming and other factors are bleaching and they end up like this". The accompanying film shows a decaying chunk of coral. Global warming is not killing these wonderful creatures, contrary to the Gore propaganda blitz. In most ocean settings coral thrives, even growing plentifully upon junked ships and oil platforms. If we want more coral, we should build more artificial reefs which are an ideal habitat for countless sea creatures. To faith based environmentalists, artificial reefs are heresy because they are not "natural". Like all living organisms, coral eventually dies. Death can occur from exposure to air, sunlight, disease, breakage, or abnormal conditions—like the El Niño (warm water) of 1998. This was the year of a very warm El Niño. The warm and cold currents of El Niño and La Niña are not connected to AGW. During this rare episode of a very warm El Niño, some Pacific coral "bleached" and died. However, since this period, coral bleached by warmed water is tough to find.

Gore warns that a doubling of CO^2 levels will acidify seawater, wiping out all the coral reefs by 2050. This is not plausible. The first known coral reefs have been found in Mesozoic sediment, when atmospheric CO^2 levels averaged rates three to six times greater than today.

Chapter 9

THE POLITICS

At the 2014 United Nations Climate Summit the UN stated that without significant cuts in emissions in all countries the window to stay within less than two degrees C of warming will soon close forever. Actually, this window will remain open for some time. A growing body of evidence suggests that the climate is much less sensitive to increases in CO_2 emissions than the IPCC assumes and that the need for reductions in such emissions is less urgent. Mounting evidence suggests that the basic assumptions about climate change are mistaken. The numbers don't add up.

According to the G10 (now the G9) they will not allow the climate to warm more than 2 degrees C. (3.6 degrees F) above pre-industrial temperatures. Since the little ice age (1400-1850), the earth's surface temperature has warmed about .8 degrees C. This leaves 1.2 degrees C (about 2.2 degrees F) to go. In one scenario of increasing CO_2 the IPCC says this threshold could be exceeded by 2040. The IPCC has consistently projected too much warming since the 1990s. The IPCC itself claims that 41% of the warming is missing. The missing heat is "hiding" in a hot spot in the sea, the atmosphere etc. After years of looking for the missing heat they have not found it, yet still insist that the Climate Models are correct and refuse to make the models reflect reality and instead have doubled down on a series of studies proving AGW based on models with no fidelity to reality.

Human caused (anthropogenic) warming depends not only on increases in greenhouse gasses but also on how "sensitive" the climate is to those increases. Climate sensitivity is defined as the global surface warming that occurs when the concentration of CO_2 doubles. If climate sensitivity is high, than we can expect substantial warming in the coming century as emissions continue to increase. If climate sensitivity is low, then future warming will be substantially lower, and it may be several generations or a thousand years before we reach what the G9 considers a dangerous level, even with high emissions.

More than a dozen observation based studies have found climate sensitivity values lower than those determined using global climate models, including recent papers published in *Environmetrics* (2012), *Nature Geoscience* (2013) and *Earth System Dynamics* (2014). These new climate sensitivity estimates add to the growing evidence that climate models are running too hot. Moreover, the estimates in these empirical studies are being borne out by the much discussed "pause" or "hiatus" in global warming. The period since 1998 during which global average surface temperatures have not significantly increased.

This pause in warming is at odds with the 2007 IPCC report, which expected warming to increase at a rate of .2 degrees C per decade in the early 21st century. The warming hiatus, combined with assessments that the climate model sensitivities are too high, raises serious questions about whether the climate model projections of 21st century temperatures are fit for making public policy decisions.

The sensitivity of the climate to increasing concentrations of carbon dioxide is a central question in the debate on the appropriate policy response to increasing carbon dioxide in the atmosphere. Climate sensitivity and estimates of its uncertainty are key inputs into the economic modes that drive cost-benefit analyses and estimates of the social cost of carbon.

Continuing to rely on climate-model warming projections based on high model-driven values of climate sensitivity skews the cost—benefit analyses and estimates of the social cost of carbon. This can bias public policy decisions. The implications of the lower values of climate sensitivity in recent studies , is that human caused warming near the end of the 21st century should be less than the 2 degrees C danger level of the G9.

This slower rate of warming, relative to climate model projections, means there is less urgency to phase out greenhouse gas emissions now, and more time to find ways and economic growth to adapt to climate change. It also allows us the flexibility to revise our policies as further information becomes available.

The sad truth is none of the energy and economic policies triggered by the demonization of CO^2 were necessary. The price paid to date is incalculable and it will go on costing far too many people, especially scientists and governments, who built their entire careers around the falsehoods.

In demonizing fossil fuels, Malthusians, wearing the green mask of environmentalists elevate their beliefs in the limits of growth over the quality of human life. Our use of fossil fuels correlates with dramatic increases in life expectancy and income, especially in the developing world. No other technology available to us today can meet the energy needs of everyone on the planet. The average American uses machine energy of 186,000 calories per day equal to that produced by 93 physical laborers, and the vast majority of this is produced by fossil fuels. If we were to give up fossil fuels and rely on the false beliefs of Malthusians and the good intentions of environmentalists, we would be plunged back into pre-industrial hell. Life expectancy would plummet and climate related deaths, especially from cold, would soar. Being forced to rely on solar, wind and biofuels would be a horror beyond anything we could imagine, as a civilization that runs on cheap, plentiful, reliable energy would see its machines dead and its productivity destroyed.

Standing atop the pyramid of Malthusian finite-resource alarmists and exploiters are those who can convert insecurities about raw materials into raw material power. As America's fracking revolution once again debunks end-of-oil myths, and oil prices plunge disproving the many predictions that increased US production would not lower gas prices because oil is priced in a global market, many wearing the green mask of environmentalism will move on to the planet's limited supply of clean air and water.

One of the many agendas behind the AGW bandwagon is self-labeled anti-capitalists. They believe the root cause of global warming is not carbon emissions, per se, but capitalism. They believe that free market economies produce too many goods for the people's own good, that is, satisfying consumer's insatiable appetites results in too much demand on the earth's finite resources. Since communist China and Russia are ruthless exploiters of the environment, this analysis is flawed. They are Malthusian's who want the political power to stop growth and lower living standards. They want to be able to force lifestyle changes via economic and regulatory controls in order to save the planet. Typical of this wing of the AGW coalition is 350.org. These Malthusians have false fronts two deep. They claim to be environmentalists, but they think they are anti-capitalists when in reality they are members of the church of Malthus all of whom have been proven wrong for 216 years.

THE CHIEF SCIENTIST OF THE UNITED STATES

The White House Office of Science and Technology Policy (OSTP) was established by Congress in 1976. Its mission is to serve as a source of scientific and technological analysis and judgment for the President with respect to major policies, plans and programs of the Federal Government. President Obama appointed John Holdren as his chief scientist.

Holdren is a classic Malthusian wearing the green mask of environmentalism. He has written:" Indeed, it has been concluded that compulsory population control laws, even including laws requiring compulsory abortion could be sustained under the existing Constitution if the population crisis became sufficiently severe as to endanger the society." Translating for our chief scientist, by "sustained under the existing constitution" he means appoint five Malthusian Supreme Court justices who believe, with the current majority in a "living constitution". In his job interview in the Senate, under questioning about his writings, Holdren recanted many of his statements. If he hadn't recanted on forced abortions etc. he never would have been confirmed.

Being a Malthusian, our chief scientist has a solution for the runaway population problem (remember the birthrate in the US is below replacement). He wrote in 1973, "Adding a sterilant to drinking water or staple foods is a suggestion that seems to horrify people more than most proposals for involuntary fertility control. To be acceptable, such a substance would have to meet some rather stiff requirements: It must be uniformly effective, despite widely varying doses received by individuals, and despite varying degrees of fertility and sensitivity among individuals"

In a 1995 paper, Holdren explained his model for sustainable development, noting "humans are included as just one species and are not treated specially". Every major religion gives mankind "dominion" over nature. Holdren and his fellow Malthusians want dominion over mankind to equate people with rats, maggots, mosquitoes, flies, and cockroaches. The next time you want to swat a mosquito, remember that, according to your chief scientist, "every species should be treated equally".

Holdren writes:" A massive campaign to restore a high quality environment in North America and to de-develop the United States must occur" By de-develop Holdren has explained he means "lower per capita energy consumption, fewer gadgets, and the abolition of planned obsolescence".

The 2014 landing on a comet supplied data refuting the theory that earth's water came from comets. Billions of years ago, the planet had very little water and then a long bombardment of asteroids heavy with ice supplied the water that covered two thirds of the Earth. In the narrow, blighted view of the finite resource Malthusian the tremendous amount of water in the asteroids doesn't exist because we live in a closed system, even as we make plans to lasso an asteroid and bring it into Earth orbit.

FRACKING, PIPELINES AND RAILROADS

Anti-carbon forces are quickly moving against the fracking revolution based on oil's threat to ground water. Better yet, the same rationale can be used by those who wish to consolidate and ratchet up collective nation state power into global governance bodies, treaties, laws and agreements. The ingenuity of entrepreneurs can slow the concentration and centralization of state power, but public sector resource protectors and allocators are every bit as persistent and ingenious.

Fracking was developed because of a loophole in the Federal Government Campaign against oil production. The loophole is called fee simple title. In almost all states, owners of private property are allowed to sell or lease the mineral rights to their own land. State and Federal regulators could not choke off the experimentation with drilling methods such as horizontal drilling. Oil companies learned to put gyroscopes on drill bit heads, and thus could drive the drill bit for miles underground looking for oil and gas deposits. The Carter administration spent 10 billion dollars on trying to develop shale oil. This was one of ten projects of similar scale that was going to alleviate the Saudi embargo on oil. All ten projects produced no useful economic energy or technology. The energy revolution was developed without any government funding, indeed against strenuous efforts by government authorities. The Malthusians since the 70s have been in full throated cry for not one more barrel of oil. Not matter how often the facts betray their false logic of finite resources they will never give up, and never stop campaigns against fossil fuels, pipelines, refineries, natural gas, oil, and coal. This campaign is always cloaked in the green mask of a false environmentalism.

Consider the environmental of pipelines vs. rail. Pipelines are safer, cheaper and cause less environmental damage than moving oil by train. Not only Keystone XL, $10 billion, but Northern Gateway $6.5 Billion, Trans Mountain $4.8 billion, Sand Piper, $2.6 billion, Constitution, $693 Billion, Line 98, $395 million, Alberta Clipper, $160 million have been stopped and or/delayed. Resistance to new pipelines and the expansion of existing ones is not slowing growth in oil output. Oil shipments by rail have increased three times since 2008. Therefore, the oil in being shipped by train causing increased environmental damage, increased risk including horrific fires, and increased costs which are passed to consumers. Malthusians do not care about the increased CO^2 emissions of trains over pipelines, so long as they can slow down growth.

Since Keystone XL was proposed, Canada's oil production has increased 39% to an average of 3.8 million barrels a day, according to Canada's National Energy Board. However, the political and regulatory snarls of Keystone XL as a template, national environmental groups are joining with local activists in a strategy aimed at prolonging environmental reviews of proposed pipelines routes and their environmental impacts. As a result, six oil and natural gas pipelines in North America costing a proposed $15 billion dollars or more and stretch more than 3,400 miles have been delayed or stopped. The purpose of the national environmental group's action is to raise the price of fossil fuels. These costs are passed to the consumer. Thus the effect of the actions is that richer people are raising the costs of poorer people. Malthusians are not trying to eliminate income inequality, although they often say they are. Their behavior reveals their preferences.

In the same period, US oil production has surged 74% to 8.8 million barrels a day, while natural gas production climbed 21% to 2.7 trillion cubic ft. in August, 2014, according to the US Information Administration. The US is now the world's largest natural gas producer and is on track to become the biggest oil producer next year. So far, from running out of the finite resources of fossil fuels, the world is awash in cheap oil and gas, spreading cheap energy around the world which is the strongest force to continue to reduce climate deaths. Not to worry, after 216 years of always being wrong about the real word effect of the finite limit of natural resources, there will continue to be more Malthusians than ever before. They just change their green, red, Club of Rome, climate change masks to hide their Malthusian agenda.

Despite rhetoric about being for all forms of energy, the Obama administration has done everything it could to inhibit fossil fuel energy production. Their strategy is to raise the price of gas, oil, coal etc. in order to support the alternatives, wind, solar etc. Their tactic is CO^2 reduction. The White has called on federal agencies to consider the climate change impact of a wide range of energy projects that require government approval. The draft guidelines by the Council on Environmental Quality affect fossil fuel projects the most, such as pipelines, terminals that export coal and liquefied natural gas, and production of oil, natural gas and coal on public lands. The draft spells out how different agencies, such as the Interior department, Federal Energy Regulatory Commission and Army Corps of Engineers, should consider the greenhouse gas emissions of projects that require environmental reviews. It also requires agencies to consider alternatives that have smaller carbon footprints. These regulations give government agencies the power to delay energy projects indefinitely. Current law requires the government to consider local environmental concerns about projects, but not global problems such as climate change. Thus, the White House can change the law via regulation without participation from Congress.

Chapter 10

SHODDY CLIMATE SCIENCE CORRUPTS PUBLIC POLICY

CO_2 is not causing dangerous warming. There is no scientific need to replace fossil fuels. Replacing them with alternative energies compounds the problems. A US Senate reports:

Comparisons of wind, solar, nuclear, natural gas and coal sources of power coming on line by 2015 show that solar power will be 173% more expensive than traditional coal power, 140% more than nuclear and natural gas power. Wind and solar's "capacity factor" or availability to supply power is around 33%, which means 67% of the time wind and solar cannot supply power and must be supplemented by a traditional energy source such as natural gas.

Tax credits have been essential to the economic viability of wind farms so far, but will not be needed with a few years. So said Christopher Flavin, now president of Worldwatch Institute in 1984. Warren Buffet, a huge investor in wind farms, said they make no sense without subsidies.

Thirty years and billions of dollars later, the wind industry is still saying it needs taxpayer support. Congress, in late 2014 debated whether to end the 22 year old production tax credit (PTC). The PTC gives wind producers a 2.3 cent tax credit expired at the end of 2013. Wind power capacity has increased by nearly 5,000% since the PTC was created and the industry now makes billions of dollars in annual revenue. Meanwhile, the credit has devolved into another example of corporate welfare.

Over the past seven years, the PTC has cost taxpayers $7.3 billion, and it is expected to pay out 2.4 billion more in 2014. Combined with other subsidies and programs, wind generators received $56.29 in government subsidies per megawatt hour in 2010, according to a 2012 report from the Institute for Energy Research. That's compared with 64 cents for natural gas and $3.14 for nuclear power.

The program operates as one of America's least known wealth distribution schemes, forcing taxpayers to pick up the tab for wind farms beyond their borders. In 2012 more than 30 states paid more in subsidies than wind farms in that state received in tax credits. Citizens in five states paid more than $100 million in taxes than they received: California ($196 million), New York, Florida, New Jersey and Ohio.

Eleven states paid into the PTC even though they have no qualifying wind production. The credit also encourages abuse, both of the electricity grid and of the taxpayer. Instead of paying wind producers based on how much of their electricity is used, the PTC pays them on how much they generate. Companies that invest in wind power thus receive tax credits to produce something that consumers may not actually want. In fact, producers often pay electricity grid operators to take their product. This phenomenon is known as "negative pricing".

Wall Street has figured out that it can use this system to its advantage. The PTC offers major corporations a chance to lower their tax rates by investing in wind energy. But investors also realize that wind farms make little financial sense if the taxpayer isn't picking up the tab.

Wind power's fluctuating growth patterns bear this out. In 1992 wind installations produced about 2.8 million megawatt hours of electricity: in 2013 wind installations produced 167.6 million megawatt hours. Yet when the PTC expired temporarily in 2000, wind installations plummeted 92% the next year. The same thing happened in 2002 and 2004, when new installations fell 76% after temporary expirations.

But the last few years deserve special mention. For most of 2012, wind producers weren't sure if the PTC would be extended at the end of the year with only 1.100 new megawatts coming on line the following year—a more than 90% drop.

Yet, Congress caved and gave the PTC a one year extension in January 2013, throwing in a bonus: Wind projects under construction by the end of the year would still be eligible for the PTC, even if they came online after the credit expired.

Corporations and wind producers, promptly rushed to cash in on the taxpayers generosity. The industry broke ground on 12,000 megawatts of new wind farms before the PTC finally expired in December. Thanks to the credit's ten year payout guarantee, tax payers still have another decade of subsidizing wind. Congress should eliminate this and all other energy subsidizes, it is time to let all energy companies compete in a fair market. This simple energy policy will provide the lowest cost, most reliable and most resource efficient energy solution.

Wind turbulence restricts the number of turbines to 5 to 8 turbines per 2.6 square kilometers. With average wind speeds of 24 mph it needs 8,500 turbines covering 2590 square kilometers to produce the power of a 1000MW conventional station. To put this in perspective, Ontario, Canada closed two 1000MW plants in 2011— the Lambton and Nanticoke coal fired plants. You need 5,180 square kilometers of land to replace them with wind power. Besides the land, you still need two 1000MW plants to back up the wind turbines. Later we will detail the thousands of birds that are killed by wind turbines every year. The birds include protected species like Bald Eagles, which wind turbines alone have a license to kill.

In California, we have a law mandating 33% of our energy must come from renewables. Governor Brown wants to go to 50%. We are at 15% and lead the nation and strains on the system are growing. We heavily subsidize wind and solar in many ways. The backbone of California electrical power is natural gas plants which supply about 64% of our base load. Because of the mandate for renewables the utilities use the renewable supply first.

This means that the natural gas plants are not operated at their most efficient and profitable rate. Consequently, our natural gas plants are not profitable to operate. This means the utilities want to close these plants, along with real estate developers, and green zealots. As we close our natural gas plants we begin to lose our base load, the ability to supply electricity when the sun is not shining and the wind is not blowing. Unless we start building more natural gas plants, which utilities do not wish to do, since they are not profitable, we will experience increasing brownouts, rationing and power interruption. California Utilities were forced to stop buying electricity from the many coal fired plants outside our borders, particularly in the Four Corners region. Many California Utilities owned portions of these plants and they were ordered to dispose of these assists and ensure that they did not continue to produce electricity from coal. Those coal fired plants still exist and are producing electricity because there is no way to enforce a California edict in Utah, Nevada, and Arizona etc. Thus there has been no reduction in CO_2 emissions, and thus no benefits from an expensive public policy.

An article titled "Wind Power is a Complete Disaster" reports the German experience:

Germany's CO_2 emissions haven't been reduced by even a single gram –and additional coal and gas-fired plants have been constructed to ensure reliable delivery. Germany seeks to get back on track to meet its energy goals after a recent increase in carbon gas emissions threatened to derail them. The government is adopting a broad catalog of measures, from new subsidies for homeowners who insulate their houses to mandatory emissions cuts for energy producers. Berlin said the steps would ensure Germany meets it target of cutting CO_2 emissions by 40% from their 1990 level by 2020—a goal twice as ambitious as the European average and one that appears increasingly beyond reach.

Germany's plan to wean the country off nuclear and fossil fuels by midcentury—a flagship policy of Chancellor Merkel dubbed the *Energiewende*, or energy revolution—made the country a model for environmentally friendly growth when it was adopted three years ago. But steps so far, including speeding up the phase-out of nuclear energy, have produced unintended consequences such as rocketing electricity pricing and rising CO_2 emissions.

Denmark has the highest percentage of wind power and their experience is telling. As the *National Post* reports:

Its electricity generation costs are among the highest is Europe.

Niels Gram of the Danish Federation of Industries says "windmills are a mistake and economically make no sense". Aase Madsen, Chair of Energy Policy in the Danish Parliament, calls it a, "terribly expensive disaster".

Because the wind can drop or surge suddenly it puts stress that can overload the grid so wind power is generally limited to 12% of the total supply. Other problems include the surge demand placed on the grid when the wind drops off, or the addition of surplus power when the electrical demand drops and wind power is still being added. A report from Britain tells of wind farms being paid to shut down turbines to prevent this problem. These are economic realities, but add in the number of birds killed, the blight on the landscape, the cost of transmission from remote locations and it is not an alternative.

There is nothing that shows the green cloak, the hypocrisy of the environmentalists who are really misguided Malthusians that push alternative energies in the name of saving the environment than bird kills at renewable energy sites.

In December 2014, an Oregon based wind energy company filed suit to block the Federal Government from releasing information on bird kills at the firm's 13 wind power sites nationwide. The plaintiff, PacifiCorp, is asking a US District Court judge in Utah to issue an injunction banning the US Department of the Interior from providing *The Associated Press* with information on how many birds are being found dead at PacifiCorp's wind installations.

In its complaint, PacifiCorp claims the information Interior would provide AP as the result of a request by the news agency under the Freedom of Information act in confidential commercial information. Warren Buffet, having correctly understood that the Keystone Pipeline would not be built, bought a railroad that is making huge profits hauling oil. The railroads are hauling so much oil, that it is crowding out agricultural products and thus raising food prices. It is more damaging to the environment to ship oil by train than by pipeline, causes huge fires, and is much more expensive. Buffet is now making a huge bet on wind power. With the mandate and subsidies for wind power, plus contracts from utilities that favor wind, sometimes paying them for wind they don't use, Buffet sees a can't lose business proposition, since Wind Power lobby has blown over resistance in congress every time the PTC comes up for renewal.

At the same time, Buffet has said that AGW proponents like Governor Cuomo are exaggerating storm damage based on his knowledge of insurance claims. There are many capitalists like Buffet who are profiting from the dislocations and high prices caused by the AGW agenda. PacifiCorp, a subsidiary of Berkshire Hathaway, is one of several wind companies that has objected to AP's coverage of bird deaths at wind energy installations.

If PacifiCorp prevails in court, you may essentially lose your right to learn about the environmental impacts of projects funded by your tax dollars on public lands owned by you. Solar companies that do business on public lands will be watching this lawsuit carefully.

Raptor kills at wind turbine farms have been observed by passersby but denied by operators. Biologists hired at wind and solar farms have been told that if they found too many dead birds they would be fired. Solar company representatives imply in the press that they document every single bird death at their facility, despite systematic surveys being done less than once a week over less than a quarter of the project footprint even at the height of the survey period.

Wind and solar power installations routinely kill birds protected by the Migratory Bird Treaty Act, by the Bald and Golden Eagle Protection Act, by the Endangered Species Act, and by a host of other state and local laws. They do so with impunity because they are protected by the green mask of Malthusians masquerading as environmentalists.

In the 1979s, the Altamont Pass Wind Resource was constructed just east of San Francisco. It is the world's largest concentration of wind mills. Some 4,500 windmills sit atop 50,000 acres producing a modest 576 megawatts of power. Equivalent natural gas plants would take up far less than 10% of the footprint, produce electricity at half the price and kill no birds.

We are deeply distressed about the continuing bird deaths, said Elizabeth Murdock, executive director of the Golden Gate Audubon Society, the chief plaintiff in the lawsuit. Nevertheless, California has a law that requires 33% of our electricity come from wind and solar. The phony greens reveal their true goal; they don't care about environmental costs in bird deaths, or other impacts to the environment. The Malthusian wants rationed, expensive, unreliable electricity generation to slow and stop growth.

Since the multitude of three bladed rotors was installed at Altamont, a significant increase in the numbers of dead birds in the area has been reported. Bird lovers demanded action. Since then lawsuits have been filed, and millions of dollars spent procuring studies to track the bird body count in an effort to determine how to address the problem. In 2008, the most extensive taxpayer funded examination was conducted by the Altamont Pass Avian Monitoring Team. It surveyed 2,500 of the turbines and kept meticulous records of the bird body count. During the study period, 1,596 rotary blade bird deaths were confirmed, including the deaths of 633 raptors. Extrapolating their data to account for all 4,500 windmills on the farm, as well as estimates regarding how many dead birds the researchers didn't count before scavengers made off with an easy meal, the monitoring team claims that in two years 8,247 birds died of turbine to the head syndrome.

The Comanche Peak nuclear power plant outside Dallas, Texas, produces about 2,300 megawatts of power, enough for 1.3 million homes. It sits on 8,000 acres and includes a large reservoir for cooling the plant, which doubles as a source of recreation. Compare that landmass to the one required for the highly publicized Pampas Wind Project, which was promoted by Dallas hedge fund manager, T. Boone Pickens. The highly touted "Pickens Plan" envisioned supplying power to an equal number of homes but required 400,000 acres of Texas real estate. Besides erecting thousands of massive masts upon which the turbines are fixed, Pickens's plan necessitated the construction of transmission towers and lines and associated service roads.

Chapter 11

Solar Power and "Streamers"

Google invested over a billion dollars in a solar facility in Nevada. The Google engineers realized that using mirrors to focus the sun light on a central tower was much more efficient than stripping electrons in solar panels. So they partnered with a utility which had crony connections with the Obama administration. This plant was supposed to deliver one megawatt of power. In reality, it operated at 25% of its rated capacity. Thus it is uneconomic to operate, even with subsidies. The utility is now seeking a bailout of another 500 million dollars. If it does not receive the bailout, it will join Solyndra on the junk heap of central planning, government funded projects.

The mirrors heat the air to over 800 degrees, which cause any birds that fly into their path to burst into flames and thus are called "streamers". To date this plant has caused 2,000 birds to perish. There is a resounding silence from the environmental community about the slaughter of birds from this plant and a new one under construction in California. The EPA, which is constantly harassing farmers and ranchers about slugs, newts, toads under the Endangered Species Act, has nothing to say about the killing of birds such as bald eagles, which has a penalty of ten years in prison.

We are subsidizing this grotesque slaughter of birds so we can pay more for an unreliable, uneconomic energy source.

The largest solar power plant of its type in the world—once promoted as a turning point in green energy—isn't producing as much energy as planned. One of the reasons is as basic as it gets. The sun isn't shining as much as expected.

Sprawling across roughly 5 square miles of federal desert near the California-Nevada border, the Ivanpah Solar Electric Generating System opened in February, with operators saying it would produce enough electricity to power a city of 140,000 homes. So far, however, the plant is producing about half of its expected annual output for 2014, according to calculations by the California Energy Commission. It had been projected to produce its full capacity for 8 hours a day, on average.

Factors such as clouds, jet contrails and weather have had a greater impact on the plant than the owners anticipated, the agency said in a statement.

It could take until 2018 for the plant backed by $1.6 billion in federal loan guarantees to hit its annual peak target, said NRG Energy Inc. which operates the plant and co-owns it with Google Inc. and Bright Source Energy. NRG sent a letter to the White House demanding that its $1.6 billion loan guarantee application be personally managed through the Energy Department just as Solyndra cost taxpayers $500 million dollars, while the crony capitalists were "protected"and never lost a dime, taxpayers may have another $1.6 billion dollar political payoff to finance.

The technology used at Ivanpah is different than the familiar photovoltaic panels commonly used for rooftop solar installations. The plant's solar thermal system, sometimes called concentrated-solar thermal, relies on nearly 350,000 computer controlled mirrors at the sight, each the size of a garage door.

The mirrors reflect sunlight to boilers atop 459 ft. towers—each taller than the Statue of Liberty. The resulting steam drives turbines to create electricity. A similar plant is under construction near Palm Springs.

The operation of such plants is highly dependent on weather conditions, and predicting when and how strongly the sun will shine is not a perfect science. Problems could include getting the thousands of mirrors pointed in precisely the right direction, especially in the cool early morning or keeping them clean in the dusty Mojave Desert.

Operators initially expected to need steam from gas-powered boilers for an hour a day during startup. After operations began, they found they needed boilers running more than four times longer, an average of 4 hours a day.

State energy regulators in August approved the plant's request to increase the natural gas it is allowed to burn by 60%. Additional natural gas could also be needed to operate boilers when clouds thicken or to maintain out-put at the end of the day and extend the capability for power production, the company said.

Because the plant requires sun-light to heat water and turn it to steam, anything that reduces the sunlight will affect steam conditions, which could damage equipment and potentially cause unsafe conditions, said the commissions which approved the request for increased gas use.

So taxpayers are at risk of paying off the $1.6 billion dollar loan guarantee, for a plant that is supposed to provide "free" electricity. When you add fossil fuels that were used to mine, manufacture and transport the mirrors, boilers, computers etc. to the gas that is burned to operate the plant does not even reduce CO^2 emissions. It is like paying your Visa bill with your MasterCard and thinking you're getting free money. For the $1.6 billion dollars they could have built four natural gas plants, each with the rated capacity of the plant, on one tenth the land. Thus they could have had four times the electricity with a tenth of the environmental damage.

Chapter 12

Solar Energy that is Good for the Economy and the Environment

There is an appropriate use for solar energy. I have seen recent examples of solar energy in remote areas of Peru and Burma. In Peru, Jose Tello, whose family received solar power panels in August 2014, said, it has changed our life style, it is much better than the candles we used, you can see a lot better.

About a third of Peru's and Burma's people have no electricity. A main problem in these regions is equipment maintenance in hard to reach areas, such as dense jungles. Solar panels installed in the rain forest would need more maintenance because of the high humidity and heat, which tends to reduce their heir efficiency and lifetime of their batteries.

A typical beginning system is a 100-watt set up. This is enough for a few light bulbs, a cellphone charger, radio and small TV. It is startling to travel to the depths of the Amazon basin, or remote Burma to see huts with a satellite antenna, TV and almost always, children watching cartoons. Although cell phone coverage is very limited, cell phones are prized possessions. One can see clearly that the Internet will arrive via smartphones in these remote regions. Hose Huaccha, 50, can now power up his cellphone at his dirt-floored home. (I am always struck by how few people in the US know where the expression dirt poor came from).

It is very difficult to enter the communities said Jessica Olivares, director of Acciona Microenergia Peru, the Spanish nonprofit that installed panels on 3,900 homes with loans from the Inter-American Development and charges a $3.40 monthly fee.

Chapter 13

ELECTRIC CARS AND THE ENVIRONMENT

Will electric vehicles lead to cleaner air and healthier people? It's a familiar back and forth: Advocates of alternative energy vehicles point to their positive environmental qualities, such as reducing carbon emissions from their tailpipe. Their opponents point out the hidden costs, such as the fact that the energy for electric cars comes largely from burning coal. About 38% of electricity comes from coal, down from well over half, since cheap natural gas has been supplanting coal. Natural gas emits about 60% less CO^2 than coal. The Sierra Club was for natural gas, before they were against it. A team lead by Christopher Tessum, an environmental engineer at the University of Minnesota, Minneapolis, set out to study the effects of human health of various forms alternative forms of various alternative ways to power a car. Their findings are in the Proceedings of the National Academy of Sciences.

The researchers investigated ten alternatives to gasoline. They include diesel, compressed natural gas, ethanol derived from corn, and ethanol derived from cellulose, and well as electric vehicles powered in six different ways by electricity from coal, natural gas, corn leaf and stalk combustion, wind, water, or solar energy.

The findings showed a dramatic swing in the positive and negative effects on health based on the type of energy used. Internal combustion running on corn ethanol and electric vehicles powered by coal were the real sinners. According to the study their health effects were 80% worse compared to gasoline vehicles. However, electric vehicles powered from natural gas, wind water or solar energy might reduce health impacts by at least 50% compared to gasoline vehicles.

"We were surprised that many alternative vehicle fuels and technology that are put forward as better for the environment than conventional gasoline vehicles did not end up causing large decreases in air quality related impacts" Tessum says.

The most important implication is that electric vehicles can cause large public health improvements, but only when paired with clean electricity. Adapting electric vehicles without taking steps to clean up electric generation would be worse for public health than continuing to use conventional gasoline vehicles.

It should be noted that there are no public health impacts from CO_2 on air quality. Environmental groups, despite the clear evidence that biofuels like ethanol add more CO_2 and are worse for the environment and public health, have continued to support public subsidies for them. California has the highest rate of electrical generation from wind and solar about 15%. California gets its base load from natural gas about 67%. There is no practical way to support an electrical grid without the use of fossil fuels, since sometimes the sun doesn't shine and the wind doesn't blow. California Rate payers are currently investing billions in trying to develop electric storage such as batteries. Electric cars, like the Tesla have $14,000 batteries which could be wired into the grid and provide some storage. However, for the foreseeable future fossil fuels will be required for the base load.

We have already experienced limitations of biofuels triggered by government subsidies. In that case it was simply US Agricultural land diverted. As one review notes: Switching to biodiesel on a large scale requires considerable use of our arable area. Even modest usages of biodiesel would consume almost all cropland in some countries in Europe. In California It will be mandated that biofuels be added to gasoline.

There is such a shortage of such biofuels, gas prices are expected to raise a $1 a gallon upon implementation. Biofuels, including ethanol, produce more CO_2 and nitrogen oxides than fossil fuels. However, advocates distract from this reality by presenting a net figure achieved by subtracting CO_2 used to grow the plant. Biodiesel has lower fuel efficiency that petrodiesel. Low temperatures are a serious limit for biodiesels. In a study of studies on whether ethanol made from corn uses more fossil fuel from tractors, fertilizer, etc. than it replaces, five studies found ethanol production used more fossil fuels than it replaced.

Geothermal has potential but is limited in location and usually far from where it is needed. The same is true for hydroelectric and tidal power. If the world really wants to solve the energy problems we need a method to reduce line loss in electrical transmission lines and a method to effectively store electricity.

Promoting energy policies based on immature science and alternative energies that are uneconomic is harming the economy. The people will pay the price and the well-connected get the subsidies and our economies will be worse off because of inefficient use of resources will cause less growth, fewer jobs, and lower incomes.

Politicians are ignoring the science and the economics. As US Senator Timothy Worth said: We've got to ride the global warming issue. Even if the theory of global warming is wrong, we will be doing the right thing. Canadian Environment Minister Christine Stewart said: No matter if the science of global warming is phony, climate change provides the greatest opportunity to bring about peace and justice in the world.

Chapter 14

MALTHUSIANS WITH GREEN MASKS

The cynical advocates of AGW are Neo-Malthusians who believe the world is overpopulated. This is the underlying theme of the Club of Rome. Parson Malthus said that mass famine and starvation are unavoidable because food production increases linearly (1, 2, 3, and 4) and population increase geometrically (2, 4, 8, and 16). Ironically this theory has failed for 216 years, yet there are more Malthusians now than ever. How many times have you heard that mankind's appetite for natural resources will outstrip nature's capacity to supply them? There have for 216 years been regular warnings that the world is running out of fossil fuels, soybeans, helium, chocolate, tungsten, lead and rare earth metals, you name it.

The latest reckoning with reality is the end of the obsession with "peak oil", which for years had serious people proclaiming that we were entering an era of permanent fossil fuel scarcity. It didn't work out that way.

That's a central lesson of 2014's dramatic fall in the price of oil, which fell to near $40 dollars a barrel of Brent crude from a June high of $112. In early Nov., when oil hovered at $80, OPEC officials warned they would intervene to hold the price at $70. However, Saudi officials failed to support a cut in output. The Saudi calculation is to drive oil prices down to slow the explosive growth of US production. Thanks to the tapping of domestic shale resources through the combination of horizontal drilling and hydraulic fracturing, US production has risen to some nine million barrels a day in 2014, from five million in 2008. Oil market firm HIS believes most oil extracted from "tight" deposits such as shale have a break-even cost between $50 and $69 dollars a barrel.

The International Energy Agency forecasts that US production will still surpass Saudi Arabia output of 9.7 million barrels a day and overtake Russia's 10.3 million, sometime in 2015. This would make America the world's largest oil producer, which it was from the dawn of the oil age through 1974. Thanks to the fracking boom, the US surpassed Russia as the world's largest natural-gas producer in 2013.

This is a nice reminder of why the Malthusian's constant warnings about finite supply have been wrong for 216 years. Technology responds to need and to price by increasing the "finite" supply.

It was the same story in the 1970s when the world responded to OPEC's embargo's by exploiting new resources in Alaska and the North Sea and again in the 1980s and 90s, when offshore drilling became technologically feasible and economically profitable at ever greater depths. And expect more where that came from, as the frackers continue to figure out how to drive down costs, and if new shale deposits in places such as Mexico, Ukraine and Argentina start to be exploited.

Also worth remembering is how spectacularly wrong some recent predictions of doom turned out to be. Nobel winning economist Paul Krugman wrote in 2010, declaring "peak oil has arrived".

What the commodity markets are telling us, Dr. Krugman averred, "Is that we are living in a finite world, in which the rapid growth of emerging economies is placing pressure on limited supplies of raw materials, pushing up their prices. And America is, for the most part, just a bystander in this story". Far from being a bystander, America has been the main oil-market innovator.

Such doom saying is that much more embarrassing because warnings of peak oil are nearly as old as the oil industry. In 1885, the state geologist said the amazing exhibition of oil was a temporary and vanishing phenomenon, one which young men will live to see come to it natural end. Given this 130 year record of predictive failure, why does the end of oil myth persist? Part of it is that peak oil is more wish than prediction, a desire to see the end of fossil fuels to serve a larger political agenda. It is also a way of scaring governments into pouring money into alternative energy sources that can't compete with oil and natural gas without subsidies and mandates. Predicting disaster can also be a profitable business and a path to speech making celebrity.

The happy ending is that the notion that the world is running out of resources always fails because of the ingenuity of entrepreneurs, spurred by necessity and incentive, and always exceeds the imagination of the doomsayers.

It is my guess that a majority of academic economists both know that Malthus has been wrong for 216 years, yet believe that "the rules have changed, and now we are going to reach the limit of natural resources". Indeed, my guess is most academics believe we live in a closed system, like a petri dish. When we reach the high glass walls at the edge of the petri dish, we die. Marx said that this idea was a slur on the working class. It is a slur on humanity, asserting we are no smarter than bacteria.

The idea that we live in a closed system is absurd, when we landed on the moon in 1969. We have robots on Mars which are sampling for methane in parts per billion. Methane is an organic chemical and a possible clue to life. After two years of sampling, one robot found a spike in the quantity of methane, much higher than the surrounding terrain. When there are 3,000 satellites in orbit, when we have had robots on Mars for years and the first test flight of the Orion, the spaceship that will take humans to Mars happened in 2014, don't we live in an open system, i.e. not a petri dish?

Elon Musk says that Mars is a great fixer upper of a planet. They are 30,000 people that have volunteered to take a one way trip to Mars. They probably just want to get away from the boring, blighted, blather of the Malthusians.

The most important factor in population growth today is the Demographic Transition Model. Demographic data show that populations decline with economic development. This is the exact opposite of what Neo-Malthusians believe. By any measure, the United States fertility rate is on the decline. In the 1950s based on registered live births, the number of children per woman was 3.75, in 2013 it was 1.86. It requires a fertility rate of 2.1 to maintain a stable population. In the spirit of the AGW warnings of catastrophes of biblical proportions, if the current rate of decline in fertility continues in 100 years there will be no babies born in the United States.

Chapter 15

I Was Born in a Cross-Fire Hurricane
The Rolling Stones

Florida is enjoying a record nine consecutive years without a hurricane making landfall. Indeed the Sandy Hurricane did not come ashore, yet the storm surge wrecked $80 billion dollars in damage if measured by federal taxpayer expense. Gov. Cuomo said this storm was a sign of the catastrophe of Climate Change. Warren Buffet, owner of many insurance companies, said based on insurance claims, Climate Change impact has been very exaggerated so far.

So what is going on? Federal and State "insurer of last resort" have shifted the economic costs of building homes on flood plains from individuals to taxpayers. Many beach homes are on their third rebuild funded by taxpayers.

Beach front and other flood plain homes are subsidized by cheap insurance both on the Federal and State level. In the talk about income inequality, we never mention that higher income home owners who own beach front are subsidized by lower income taxpayers.

In Florida the state owned, Citizens Property Insurance Corporation is a good example. The subsidy to insurance price incentives building homes on beaches making the ultimate costs to taxpayers every higher. The fortuitous run of hurricane-free years (which went against every AGW prediction that warmer water would drive more hurricanes) has allowed the state to keep insurance costs artificially low for many homeowners and still build up coffers to pay claims. Yet, a devastating storm or series of smaller ones risk wiping out the cushion built up by Citizens Property Insurance.

It's not a strategy, its dumb luck. Taxpayers fail to grasp the potential costs of this cheap subsidized insurance now and being on the hook for tens of billions of dollars later.

The current insurance dynamics have their roots in the aftermath of a 2004-2006 spate of some of the costliest storms in US history—including Wilma, Ivan and Charlie—when state regulators refused to allow the industry to raise rates. Many private insurers pared their exposure sharply amid the subsequent quiet spell. For example, Allstate Corp's local affiliate, now called Castle Key Insurance Co., had over 500,000 policies in force in 2006. Today they have around 200,000 policies, mainly in less risky areas of the state that can bear market rates.

This is an age old strategy of government central planners. They create a shortage with price controls, and then offer a government solution to the problem they created. The solution causes a crisis, which has to be solved by more taxpayer's dollars and more regulation. Hayek called this the "Fatal Conceit", the confidence of Government Central planners that they can design a solution to any problem. Indeed the proponents of AGW suffer from the ultimate fatal conceit, the confidence that they can control the temperature of the planet, by controlling a single variable, CO^2.

But wait there's more. Based on historical patterns going back hundreds of years, there is a two-thirds chance of a hurricane making landfall in Florida in any given year. Homeowners in Southeastern counties are more likely to have Citizens (state owned) as their insurer and also most likely to suffer storm damage. This when Citizens needs the inevitable bailout, the relatively poorer rest of Florida will bail out richer Southeaster Miami-Dade County.

Florida's Insurance Consumer Advocate pegs the cost of a 1 in 100 year storm at $53 billion dollars. A storm like the 1926 Category 4 hurricane that hit Miami might do $125 billion dollars in damage today, according to risk consulting firm Karen Clark and Co.

What is more, even these forecasts assume a single awful storm and up to 30 years to pay back excess claims. But, just as Florida has enjoyed a nine year quiet period, it has faced as many as 14 consecutive years with storms making landfall and many years with multiple hits.

When bailout day arrives, what politician would choose to explain to citizens that they had been had to the tune of billions of dollars by another central planning scheme gone wrong, when they can cry out "Climate Change" and demand, a la Hurricane Sandy, more billions from Federal taxpayers.

1950, a cool year, holds the record for the most powerful storms, with eight major hurricanes. Trying to pin the intensity and frequency of storms on global warming is folly. Like all kinds of weather, and climate is average weather, hurricanes just happen. On average, close to seven hurricanes every four years stride the US. Two major hurricanes cross the US coast every three years.

Consider these hurricanes, none of which occurred in particularly hot years:

Deadliest Hurricane: More than 8,000 people perished in 1900, when a Category 4 hurricane barreled into Galveston, Texas. The storm surge exceeded 15 ft., and winds blew at 130 mph, destroying more than half the cities homes.

Most Intense hurricane: An unnamed slammed into the Florida Keys during Labor Day, 1935. Researchers estimated sustained winds reached 150-200 mph with higher gusts. The storm killed an estimated 408 people.

Greatest Storm Surge: In 1969, Hurricane Camille produced a 25 ft. storm surge in Mississippi. Camile, a Category 5 storm, was the strongest storm of any kind ever to strike mainland America. When the eye hit Mississippi, winds gusted up to 143 people along the coast of Mississippi and led to another 113 deaths as the weakening storm moved inland.

Compare the 8,000 deaths from the deadliest hurricane in 1990 with the 250,000 deaths caused by the Tsunami in Thailand in 2012. It is pretty hard to blame Tsunamis on global warming, since they are caused by earthquakes.

Hurricanes, one of the favorite proofs that advocates of anthropogenic global warming use to validate their claims, have become a big bogeyman of climate change. There were demonstrations outside government hurricane forecaster's headquarters, because they weren't forecasting enough hurricanes.

Reporters can write scary stories, senators and governers can make unfounded pronouncements, but the only consensus regarding the connection of hurricanes and global warming is that there is no connection between the two.

Nevertheless, here is Al Gore: "The trend toward more Category 5 storms—the larger ones—and the trend toward stronger and more destructive storms appears to be linked to global warming and specifically the impact of global warming on higher ocean temperatures—and makes them more powerful". The most common tool of propaganda is repetition and this myth of global warming causing hurricanes has been repeated so many times it is accepted as gospel by the faithful.

In May 2008, the country of Myanmar (Burma) was in the direct path of a cyclone (hurricanes in the southern hemisphere are known as cyclones). A Category 3, on a scale of 1 to 5, the cyclone had winds of 120 mph when it made landfall, and surged ocean water into the low-lying, heavily populated coast for many miles. Even though the path of the storm had been accurately forecast for many days, an effective notification system had not been instituted and the death toll is estimated at 100,000. Myanmar has always had cyclones and many had been bigger than Category 3.

On May 6, on NPR, Al Gore said: "As we're talking today, the death count in Myanmar from the cyclone that hit there yesterday has been rising—and we are seeing the consequences that scientists have long predicted might be associated with continuing global warming".

If you want to observe a branch of meteorology, where there is virtual consensus regarding global warming, look no farther than those who actually study and forecast hurricanes and cyclones. These scientists know Gore and the AGW faithful are unconstrained by facts or knowledge.

Stanley Goldberg of the National Oceanic and Atmospheric Administration's Hurricane Research Division, said: "It is a blatant lie put forth in the media that makes it seem there is only a fringe of scientists who don't buy into anthropogenic global warming".

Dr. Ivar Giaver, Noble Prize winner for Physics stated:" I am a skeptic—global warming has become a new religion". Joanne Simpson, the first woman in the world to receive a PhD in Meteorology, author of more than 190 scientific studies, "Since I am no longer affiliated with any organization nor receiving any funding, I can speak quite frankly. As a scientist, I remain skeptical—the main basis of the claim that man's release of greenhouse gases is the cause of the warming is based almost entirely upon climate models. We all know the frailty of models concerning the air-surface system."

Dr. Harrison Schmidt, (Apollo 17, 1972, one of the dozen people who have walked on the Moon) in 2009, resigned from the Planetary Society that was founded by Carl Sagan. The society's main mission is "to inspire the people of Earth to explore other worlds, understand our own and seek life elsewhere." This society has fallen into the Malthusian trap of finite resources in a closed system and use AGW to control growth. Unable to change the society, Schmidt who in the 1970s also served as a US Senator, resigned with a statement.

"As a geologist, I love Earth observations. But it is ridiculous to tie this objective to a "consensus" that humans are causing global warming when human experience, geologic data and history, and current cooling can argue otherwise. Consensus as many have said merely represents the absence of definitive science. You know as well as I, the "global warming scare" is being used as a political tool to increase government control over American lives, incomes and decision making. It has no place in the Societies activities."

Chapter 16

CURRENT CLIMATE MODELS CAN NOT WORK

IPCC climate models are the vehicles of mass deception for the AGW claim that human CO_2 is causing global warming. They create the results they are designed to produce. GIGO, (garbage in, garbage out) reflects that most people working with computer models know the problem. In climate science, it actually stands for Gospel In, Gospel Out. The Gospel Out results are the IPCC predictions (projections), and they are consistently wrong. It was computer modelers who dominated the Climatic Research Unit (CRU), and through them, the IPCC. Society is still awed by computers, so they attain an aura of accuracy and truth that is unjustified.

A Climate Model is only as good as the structure which it is built, the weather records. The gap between the data and the climate models began at the CRU. The founder of the CRU, H. H. Lamb, wrote in his autobiography, "It was clear that the first and greatest need was to establish the facts of the past record of the natural climate in times before any side effect of human activities could be important".

One of Lamb's most important studies established The Medieval Warming Period (MWP). Temperatures in the MWP were higher than they are today. The MWP was followed by the Little Ice Age. These studies showed that there was natural climate change before the Industrial Age. These studies became the first target of the proponents of AGW. Michael Mann introduced the infamous "hockey stick". He used tree rings as proxies for temperature and claimed that there was no MWP. He claimed the tree rings showed that temperatures were low and stable until the start of the Industrial age and that the rapid

IPCC 1990, Figure 7c

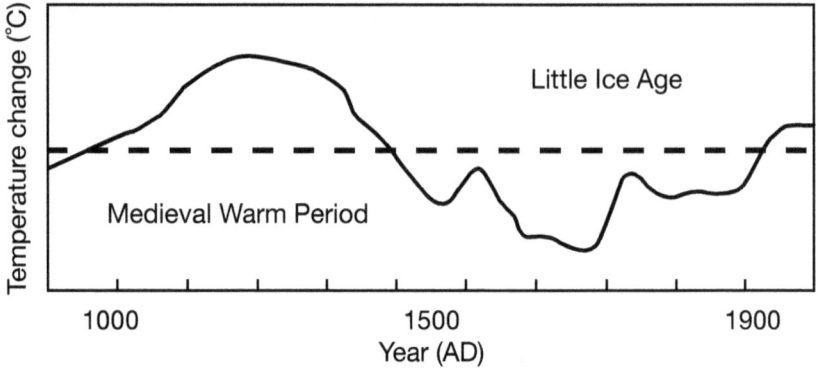

IPCC 1991

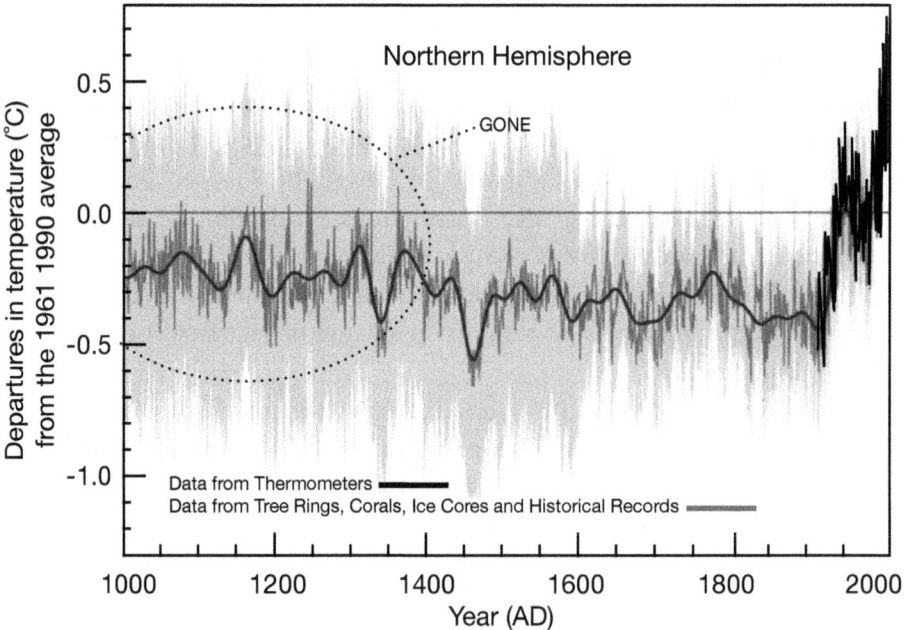

"Climategate" 2009

```
;
; Apply a VERY ARTIFICAL correction for decline!!
;
yrloc=[1400,findgen(19)*5.+1904]
valadj=[0.,0.,0.,0.,0.,-0.1,-0.25,-0.3,0.,-
0.1,0.3,0.8,1.2,1.7,2.5,2.6,2.6,
2.6,2.6,2.6]*0.75   ; fudge factor
if n_elements(yrloc) ne n_elements(valadj) then message,'Oooops!'

yearlyadj=interpol(valadj,yrloc,timey)
```

rise of atmospheric CO_2 caused record high temperatures. In leaked emails called "Climategate" he admitted he used a "trick" to hide the decline of temperatures in the 20th century by dropping the tree ring data and substituting real temperatures. He and his followers, (he has co-authored papers by 43 climate scientists, still claim the hockey stick is correct.

Lamb wrote about his successor at the CRU, Tom Wigley, "after a few years almost all the work on historical reconstruction of past climate and weather situations, which first made the unit well known, was abandoned".

Wigley promoted the application of computer models, but Lamb knew they were only as good as the data used for their construction. Lamb is still correct. The models are built on data, which either doesn't exist, or is by all measures inadequate.

Climate Models are based on grids on the surface of the Earth 3 degrees by 3 degrees. There are so few stations of adequate length or reliability, the mathematical formula for each grid cannot be accurate. Vast areas of the world have no weather stations at all. About 70% of the grids have no data at all.

Only 100 stations have records of 100 years or more and almost all of them are in heavily populated areas of the US or Western Europe and subject to urban heat island effect (UHIE). Satellites began to supply temperature data in 1979. The surface record remained the standard for the IPCC Reports.

For example, a report from 1954 to 2003 shows "no data" for the Arctic Ocean which is almost the size of Russia. In order to deal with this lack of data, the modelers use a technique called Parametization. They make an assumption that the temperature of any weather station is used for a circle with a diameter of 1200 km. Imagine a weather station in Santa Barbara where the temperature is 70 degrees. When there is no other weather station in that circle, the temperature of Santa Barbara is used for San Francisco, Las Vegas, San Diego, and hundreds of miles into the Pacific Ocean.

Empirical Test of Temperature Data

A group carrying out a mapping project, trying to use data for practical application, confronted the inadequacy of the temperature record.

"The story of this project begins with coffee, we wanted to make maps that showed where in the world coffee grows best, and where is goes after it has been harvested. We explored worldwide coffee production data and discussed how to map the optimal growing regions based on the key environmental conditions: temperature, precipitation, altitude, sunlight, and wind and soil quality.

The first extensive dataset we could find contained temperature data from NOAA's National Climatic Data Center. So we set out to draw a map of the Earth based on historical monthly temperature. The dataset includes measurements as far back as the year 1701 from over 7,200 weather stations around the world.

Each climate station could be placed at a specific point on the globe by their geospatial coordinates. North America and Europe were densely packed with points, while South America, Africa, and East Asia were sparsely covered. The list of stations varied from year to year, with some stations coming online and other disappearing. That meant that you couldn't simply plot the temperature for a specific location over time. We couldn't find the other data relating to precipitation sunlight and wind."

Anthony Watts's research showed that the US record has only 7.9% of weather stations with a less than 1 degree C accuracy.

Precipitation Data a Bigger Problem

Water, in all its phases, is critical to movement of energy through the atmosphere. Transfer of surplus energy from the tropics to offset deficits in Polar Regions is largely in the form of latent heat. Precipitation is just one measure of this crucial variable. It is very difficult to measure accurately, and records are completely inadequate in space and time. An example of the problem was exposed in attempts to use computer models to predict the African monsoon. Alexander Giannini, a climate scientist at Columbia University, "Some models predict a wetter future, other a drier one. They cannot all be right".

PEAK GLOBAL TEMPERATURE IN 1998

After 1998 CO_2 levels increased but global temperatures leveled and declined. IPCC projections were wrong. According to their hypothesis this could not happen. They failed as renowned physicist Richard Feynman explains;

"In general, we look for a new law by the following process: First we guess it, the we compute the consequences of the guess to see what would be implied if this law that we guessed is right; then we compare the result of the computation with the result of the computation to nature, with experiment or experience, compare it directly with observation, to see if it works. If it disagrees with experiment, it is wrong. In that simple statement is the key to science. It does not make any difference how beautiful your guess is, it does not make any difference how smart you are, who makes the guess, or what his name is—if it disagrees with experiment, it is wrong."

The IPCC acknowledged the problem in their Report (AR5) released in September 2013, but ignored it, as usual. AR5 concludes:

Globally, CO_2 is the strongest driver of climate change compared to other changes in the atmospheric composition, and changes in surface conditions. Its relative contribution has further increased since the 1980s and by far outweighs the contributions from natural drivers.

This is only true because they did not consider most natural drivers. They created unreal explanations, ignored contradictory evidence; used computer model generated data as real data in other computer models and used theoretical idea as real. They made it up as they went along. The also moved the goalposts. Since 1998 the temperature decline has continued in a natural pattern related to declining solar activity. The trouble was CO_2 levels continued to increase in direct contradiction to their hypothesis. According to the IPCC 90 plus certainty claim, this cannot happen. It is a perfect example of T. H. Huxley's observation, "The great tragedy of science—the slaying of a beautiful hypothesis by an ugly fact." When Einstein was told that the Germans had tasked 300 scientists with disproving his Relativity Theory, Einstein said," they don't need 300 scientists, they just need one fact".

Proper Science would go back and reconsider the hypothesis. Instead the alarmists just moved the goal posts by changing the terminology. What was global warming caused by CO^2, became climate change.

Climate is just average weather. Climate Change is a term of art, propaganda which implies that every time the weather changes this is a proof of APG.

The IPCC had already switched from predictions to scenarios. These included low, medium and high temperature projections primarily determined by increasing industrial growth and the associate production of CO^2. The three projections of temperature are all too high compared to reality, the ugly fact which is killing AGW, month by month, year by year. Temperatures fell a half degree from 1944 to 1972 or 28 years. Temperatures rose from 1972 to 1998 or 26 years. Temperatures have failed to rise from 1998 to the present. If temperatures still aren't climbing in a few more years AGW will be deader than disco. Don't worry, the Neo-Malthusians wearing the green mask, will come up with different scary scenarios wearing a different mask. How about a zombie apocalypse? That's catching on.

Chapter 17

WHY AGW PREDICTIONS OVERSTATE TEMPERATURE PREDICTIONS (PROJECTIONS)

We have predictions or projections of temperature increases on the public record since 1989. The IPCC insists that their temperature projections are not predictions. The likely reason for this is that their projections, including Hansen's predictions in congressional testimony in 1989 have all overstated future temperatures. Hansen's testimony in 1989 included three scenarios, in the best case the world would stop using any new fossil fuels by 2005. Even in this case, his prediction grossly overstated temperatures. The IPCC has followed this same pattern of grossly overstating temperature increase. We have 26 years of failure and the IPCC keeps increasing their level of confidence in their projections.

Someone said economists try to predict the tide by measuring one wave. The IPCC essentially try to predict (project) the global temperature by measuring one variable. The IPCC compound their problems by projecting the temperature variable with the influence of the economic variable.

Use of circular arguments is standard operating procedure for the IPCC. For example, they assume a CO_2 increase causes a temperature increase. They then create a model with that assumption and when the model output shows a temperature increase they claim it proves their assumption.

They double down on this by combining an economic model that projects a CO_2 increase with their climate model projection. To make is look more accurate and reasonable they create scenarios based on their estimates of future development. This creates what they want, namely that CO^2 will increase and temperature will increase catastrophically unless we shut down fossil fuel economies very quickly.

All their projections failed, even the lowest as, CO^2 continued to rise and global temperatures "paused". Instead of admitting their work and assumptions were wrong, they scramble to blur, obfuscate and counterattack.

The IPCC claim 95% certainty about their climate science and presumably about their projections. The problem is all were wrong from the start. As early as the 1995 report they had switched to projections. They gave a range of projections or scenarios from low to high, but even the lowest was incorrect. Roger Pielke Jr., explained the assumptions for the scenarios were unrealistic, especially about the technological progress in energy use and supply.

Most people assume the projections are solely a function of the climate science and climate models, but that is not the case. The Richard Lindzen, the MIT professor of meteorology said in an interview with James Glassman in the 2001 IPCC report, "was very much a children's exercise of what might possibly happen" prepared by a "peculiar group" with "no technical competence". However, it achieved their political objection of isolating and demonizing CO^2.

After the release of the 2001 Third Assessment Report (TAR) two papers by Ian Castles and David Hendersen (C&H) were published drawing attention to the problems with the emission scenarios used to produce the three projections, Castles explained the concerns as follows: "During the last three years I and a co-author (David Henderson, for Head of Economics and Statistics of the Dept. of Economics and Statistics at OECD) have criticized the IPCC's treatment of economic issues.

Our main single criticism has been the panels use of exchange rate converters to the GDP's of different countries onto a common basis for estimating and projecting output, income, energy intensity etc. This is not approved (because it is highly inaccurate and misleading) under the internationally agreed System of National Accounts which was unanimously approved by the UN Statistical Commission in 1993, and published later that year by the United Nations, the World Bank, the IMF, the OECD, under cover of a Foreword which was personally sighed by the Heads of the five organizations.

As one commentator noted:

These two economists have shown that the calculations carried out by the IPCC concerning per capita income, economic growth and greenhouse gas emissions in different regions are fundamentally flawed, and substantially overstate the likely growth in developing countries. (The report said that developing countries would grow so fast that they would produce more CO^2 than developed countries) The results are therefore unsuitable as a starting point for the next IPCC assessment report, which is due to be published in 2007. Unfortunately, this is precisely how the IPCC now intends to use its projections.

The IPCC response was typical of the arrogant superiority and belief in there unassailability that pervades most of their dealings:
On December 8, 2003, at the Milan COP9 Dr. Pachauri release a press statement which criticized the arguments which Castles and Hendersen have been making in this debate.

Pachauri's charges against C&H were false personal attacks. Richard Tol commented on C&H and the IPCC response:

C&H criticized the IPCC for using market exchange rates in the economic accounting used as a basis for its SRES scenarios. This started as a technical dispute. However, the initial IPCC response, which can be characterized as "We are the IPCC, We do not make mistakes. Please go away. He raised the stakes and turned the debate into one about the credibility of the entire IPCC, a debate that now includes politicians and the public. Howard Herzog of MIT recently summarized this as the "IPCC is a four letter word".

Under the IPCC's B1 scenario, in the year 2100 all of the following countries are projected to have higher GDP's than the US: Germany, Italy, France, Japan, the Baltic States, South Korea, North Korea, Malaysia, Singapore, Hong Kong, South Africa, Libya, Algeria, Tunisia, Saudi Arabia, Israel, Turkey and Argentina.

All of the IPCC's predictions (scenarios) have failed and are pathetic, this one is just plain hilarious. North Korea will have a higher GDP than the United States in 2100? Do you think this could be an honest mistake or just a way to exaggerate future CO_2 in order to exaggerate future temperatures? This is just more fudge from the IPCC climate alarmists fudge factory.

Chapter 18

SEA LEVEL RISE & CLIMATE REFUGEES

In 2005, the United Nations Environment Program (UNEP) warned that imminent sea level rise, increased hurricanes and desertification caused by AGW would lead to massive population disruptions. The organization highlighted areas that were supposed to be particularly vulnerable in terms of producing climate refugees. Especially at risk were regions like the Caribbean and low laying Pacific islands and coastal areas.

The UNEP predictions in 2005 claimed that by 2010, some 50 million climate refugees would be frantically fleeing from those regions. However, not only did the areas in question fail to produce a single climate refugee, by 2010 population levels for these regions were actually rising. In many cases, the areas that were supposed to produce waves of climate refugees and becoming uninhabitable turned out to be some of the fastest growing places on Earth.

In the Bahamas, for example, according to the 2010 census, there was a major increase in population, going from around 300,000 in 2000 to more than 350,000 in 2010. The population of St. Lucia, meanwhile grows by about 5% during this period. The Seychelles grew by about 25%.

In China, the top six fastest growing cities were all within the areas highlighted by the UN as likely sources of climate refugees. Many of the fastest growing cities in the US were also within the climate refugee danger zone.

Rather than apologizing for its mistake after being first exposed by Gavin Atkins at Asian Correspondent the UN with an Orwellian cover-up seeking to erase all evidence of its fatally flawed prediction. First the UNEP took it's "climate refugees" map down from the Web.

Then it tried to distance itself from the outlandish claims, despite the fact the map was created by a UNEP cartographer, released by the UNEP, and repeatedly hyped by UNEP in its scaremongering campaign. Eventually, as more media around the world began picking up the story, the UN agency claimed the map was removed because it was "causing confusion".

On June 30, 1989, *The Associated Press* ran an article headlined: "UN official predicts disaster, says greenhouse effect could wipe some Nations off map", In the article, the Director of the UNEP's New York office was quoted as saying "entire nations could be wiped off the face of the Earth by rising sea levels if global warming is not reversed by the year 2000." He also predicted "coastal flooding and crop failures" that "would create an exodus of eco-refugees.

In the final 2007 report, widely considered the "gospel" of "settled science", the UN IPCC predicted the Himalayan glaciers would melt by 2035 or sooner. It turns out this "settled science" was lifted from a World Wildlife Fund propaganda piece. The IPCC recanted the claim after initially defending it.

Chapter 19

NEVER WATCH SAUSAGE, POLITICS OR UN CLIMATE SCIENCE BEING MADE

The UNFCCC predetermined the results of the IPCC work by directing them to study only human causes of climate change. The IPCC then narrowed the focus to human produced CO^2 as the cause of warming. They directed their efforts to proving rather than disproving their hypothesis. Central to this objective was the need to have atmospheric CO^2 levels rise constantly because of a constant rise in human production of CO^2.

The IPCC controlled results of rising atmospheric levels with data from warming advocate Charles Keeling's and later his son Roger's, measurements at Mauna Loa. There is a fascinating, but disturbing correspondence on this issue between Ernst Georg Beck and Roger Keeling. Beck had to be dismissed because his work showed that 19th century levels of atmospheric CO^2 were much higher than used by the IPCC and created by Guy Calendar and Tom Wigley. The IPCC controlled the history of the level of CO^2 and production of the annual increase in human production of CO^2.

In their 2001 Report the IPCC note the increase of CO^2 from 5.5 GtC. (gigatons of carbon) human sources to 7.5 GtC. In the FAQ section they answer the question "How does the IPCC produce its inventory guidelines?" as follows:

Utilizing IPCC procedures, nominated experts from around the world draft the reports that are then extensively reviewed twice before approval by the IPCC.

In a 2008 article, Castles notes about the 2007 report:

Unfortunately, the assumptions it uses overstate potential man-made global warming by a large measure.

In 2001 IPCC based its predictions of substantially warming temperatures during the next century largely on forecasts of explosive growth in Third World economies—and hence emissions—during the twenty-first century. The panel actually predicted Third World nations would grow so fast they would surpass the economies of wealthy Western nations.

Economists pointed out the unrealistic assumptions, but in the six years since these IPCC gaffes, little appears to have changed.

Richard Tol commented on the changes for AR5:

IPCC AR5 of Working Group 1 will therefore by based on scenarios-formerly-known-as-SRES. They're now called RCP (Representative Concentration Pathways).

William Kininmonth, former head of Australia's National Climate Centre and their delegate to the WMO Commission for Climatology and author of the insightful book, *Climate Change: A Natural Hazard*, wrote the following:

I was at first confused to see the RCP concept emerge in AR5. I have come to the conclusion that RCP is no more than a sleight of hand to confuse readers and hide absurdities in the previous approach.

You will recall that the previous carbon emission scenarios were supposed to be based on solid economic models. However this basis was challenged by reputable economists and the IPCC economic modeling was left rather ragged and a huge question mark hanging over it.

I sense the RCP approach is to bypass the fraught economic modeling: prescribed radiation forcing pathways are fed into the climate models to give future temperature rise, if the radiation forcing plateaus at 8.5 W/m2 sometime after 2100 then the global temperature rise will be 3C. But what does 8.5 W/m2 mean? To reach a radiation forcing of 7.4 W/m2 would thus require a doubling again, 4 time CO_2 concentration. Thus to follow RCP8.5 it is necessary for the atmospheric concentration equivalent to exceed 1120 ppm after 2100.

We are left questioning the realism of a RCP 8.5 scenario. Is there any likelihood of the atmospheric CO_2 reaching 1190 by 2100? IPCC has raised a straw man scenario to give a "dangerous" global temperature rise of about 3C early in the 22nd century knowing full well that such a concentration has an extremely low probability of being achieved. But, of course, this is not explained to the politicians and policy makers. They are told of the dangerous outcome if the RCP8.5 is followed without being told of the low probability of it occurring."

The underpinning of the climate science and economics depends on accurate data and knowledge of mechanisms. We know there was insufficient weather data on which to construct climate models and the situation deteriorated as they eliminated weather stations and adjusted then cherry picked data. We know knowledge of mechanisms is inadequate because the IPCC WGI Science Report says so;

"Unfortunately, the total surface heat and water fluxes are not well observed."

Or

"For models to simulate accurately the seasonally varying pattern of precipitation, they must correctly simulate a number of processes (e.g. evapotranspiration, condensation, transport) that are difficult to evaluate at a global scale.

In a perverse way the IPCC acknowledge this with their attempt to claim the "pause" in temperatures for the last 18 years was due to some "deep ocean" process. Again Kininmonth acutely observes the comment in the SPM that:

"There may also be a contribution from forcing inadequacies and, in some models, an overestimate of the response to increasing greenhouse gas and other anthropogenic forcing.

Economic projections are even more difficult because of lack of data, an inability to anticipate public feedback and political reaction and the impossibility of anticipating technology and innovation.

The IPCC has already shown that being wrong or being caught doesn't matter because the cause the objectives of the scary headline is achieved by the complete disconnect between their Science Reports and the Summary for Policymakers (SMP). It is also no coincidence that the SPM is released before national politicians meet to set their budgets for Climate Change and the IPCC. As Saul Alinsky insisted in rules for radicals the end justifies the means.

The IPCC participants are chosen by the national weather agencies of each member of the World Meteorological Organization (WMO). The IPCC required people who would achieve the political and scientific objective of identifying human activities as the cause of global warming, and later climate change, generally referred to as Anthropogenic Global Warming Theory (AGW). Their work effectively thwarted the standard scientific method of disproving the theory. Scientists who dared to question the theory were derisively called skeptics. When this epithet didn't stop them, they were called deniers with its holocaust connotations. Most of the so-called skeptics were well qualified but excluded from the IPCC, making it a carefully selected group.

Some, such as Richard Lindzen and Alfred P. Sloan, professor of meteorology at MIT, participated—hoping to have a reasonable scientific input but eventually gave up. There is little doubt, Lindzen said, that the IPCC process has become politicized to the point of uselessness.

How did the IPCC maintain control and convince many, including political leaders, they were right and were the authority? Beyond using UN agencies as vehicles, they had the challenge of running an apparently open process while keeping total control.

The controlled who participate and who were the lead authors, especially of critical chapters. Richard Lindzen notes:

IPCC's emphasis, however, isn't on getting qualified scientists, but on getting representatives from over 100 countries—the truth is only a handful of countries do quality climate research. Most of the so-called experts served merely to pad the numbers.

They published the political document, the Summary of Policymakers (SPM) before the Technical (Science) Report of Working group I was issued, and making sure the Technical Report matched the SPM.

Lindzen again: "The IPCC clearly uses the Summary for Policymakers to misrepresent what is in the report. They used wording in the SPM to catch the media and public attention. It is difficult to describe scientific information for an essentially non-scientific audience through the media; what one blogger describes as the Math Free Zone of Journalism". Using non-mathematic terminology in the SPM, exemplified by the labels set out in a table in the third report, such as Very unlikely (1-10%) Likely (66-90%) and Very likely (90-99%). The percentages are not used in the technical report. As one study says,

"How the assessment frames the information is determined by the choices and goals of the users. For the IPCC, this includes focusing on negative impacts of warming when there are positive effects and including and highlighting studies that appeared to identify a "human signal".

Here is Lindzen's summary of the IPCC process:

It uses summaries to misrepresent what scientists say; uses language that means different things to scientists and laymen; exploits public ignorance over quantitative matters; exploits what scientists can agree on while ignoring disagreements to support the global warming agenda; and exaggerates scientific accuracy and certainty and the authority of undistinguished scientists.

How the IPCC Distorted the Historical Record

A major reason the deception was easy involved public lack of knowledge of the extent and nature of climate change. There are three segments of temperature data. The boundaries between temperature periods are the result of technical or methodological measures not natural factors. The instrumental record that covers just over 100 years, the historical record of human observations covering about 3,000 years and the rest of time is in the biologic or geologic time. Most records, especially in the historical period, are called proxy records, that is, they are a secondary indication of weather and climate such as the first arrival of geese in the spring or the date of cherry blossoms in Japan.

All records contradict claims made by official climate science. They show:

Temperatures vary considerably and in very short time periods.

Global temperatures were much warmer than today on many occasions.

Temperature increases precede CO^2 increases.

Current climate changes are not unprecedented.

The Antarctic ice core record covering 420,000 years appeared in 1991. It shows how much temperatures vary with a range of 12°C.

This became sidelined by the fact they also published the varying CO^2 over the same time period. This relationship was immediately seized and promoted as evidence that CO^2 was driving temperature. Only a few years later we learned that the relationship is exactly the opposite. We now know that in every record of any duration for any period, temperature increases before CO^2. This completely contradicts and therefore negates the basic assumption of the AGW hypothesis. In the late 1990s when the Al Gore's film, *An Inconvenient Truth,* was made, it was well known that in the ice core data featured in the film the temperature started up an average of 800 years before the CO^2. The mechanism of the correlation between temperature and CO^2 is probably that increasing temperatures over hundreds of years force CO^2 out of the oceans and into the atmosphere.

Nevertheless, inconvenient truth and the big lie that the ice core data became standard instruction in many public schools. In Britain, where they have a law against propaganda in public schools, a parent sued the school his child was in, claiming the film was propaganda. A Judge, after reviewing the 14 major errors in the film, ruled that it was propaganda. To this day AGW advocates will deny the obvious, the fact that even though the temperature rises 800 years before the CO^2, they will insist that the CO^2 causes the temperature increase. The spin they usually use is that a mysterious factor x starts the temperature up for 800 plus years and then CO^2 makes the temperature rise. The more sophisticated defense by the AGW advocates is that with Industrialization the rules changed.

The data show that the rules have not changed. Rising CO^2 does not cause rising temperature, since temperatures fell by .5 degrees F from 1944 to 1970, while CO^2 was rising rapidly after WWII. There has been no warming trend since 1998 while CO^2 was rising. The rules have not changed and there is no statistical correlation between rising CO^2 and rising temperature. The EPA knows this and refuses to release the public information which includes the correlation, called the R2. The EPA says that this public information is only for use by experts.

Despite limitations, such as a 70 year smoothing average, ice core records provide another contradiction to a major IPCC claim, namely that the world is warmer than ever before.

A graph from the Greenland GISP 2 ICE Core shows how much temperature changes naturally with dramatic swings over a range of 12 C. The world was warmer than today for most of the last 10,000 years: a period known as the Holocene Optimum. It is recorded in the Greenland Ice cores.

It was 3 degrees C warmer on average than today for most of the 10,000 years. All the species, including polar bears, said to be endangered today by climate change survived these 10,000 years. This ice core temperature record also show the Medieval Warm period (MWP) which happened about 1,000 years ago. Note the world was warmer today for most of the preceding 8000 years. This gives lie to claims about the demise of the endangered polar bear because they survived these conditions.

These records are valuable because they transcend the boundaries. Tree ring records are useful but are vulnerable to being exploited because they are easily distorted. Tree rings are a source of data used to span the instrumental and history stages of reconstruction. The original use, called dendrochronology, determined age of the tree by counting annual growth rings. Then, with the work of people like A.E. Douglas, they became valuable reconstructing solar activity as they registered variation carbon 14 in the atmosphere. The use for the IPCC record was totally inappropriate.

They used tree rings for climate studies with the assumption that they indicate temperature change. In reality, they reflect the growth pattern and are the result of a multitude of environmental factors. Temperature is a minor factor. Precipitation is the main determinate of growth, as any gardener knows.

Despite this a few, especially those associated with the Climate Research Unit (CRU) of East Anglia, began producing studies using tree rings solely as an indicator of temperature. The pivotal paper published in *Nature* in 1998 by Mann, Bradley and Hughes titled, *Global-scale temperature patterns and climate forcing over the past six centuries*, became known as MBH98. Mann was the principal author, so John Daly gave him credit and wrote, "Mann completely redrew the history, turning the Medieval Warm Period and Little Ice Age into non-events, consigned to a kind of Orwellian "memory hole". The tree ring data he produced formed the handle of a stick. Then, using an inappropriate technique he tacked on the modern instrumental record to form the blade. We later learned this was necessary because the tree ring showed declining temperatures for the 20th century. It was terrible science and statistically wrong. Despite this, the paper passed peer review.

The hockey stick graph, remarkably quickly, became the orthodoxy. As John Day explained: What is disquieting about the Hockey Stick is not Mann's presentation of it originally. As with any paper, it would sink into oblivion if found to be flawed in any way. Rather it was the reaction of the greenhouse industry to it—the chorus of approval, the complete lack of critical evaluation of the theory, the blind acceptance of evidence which was so flimsy. The industry embraced the theory for one reason and one reason only—it told them exactly what they wanted to hear.

Mann became the lead author of the chapter Observed Climate Variability and Change in the 2001 IPCC Report. He was also a contributing author on other chapters. The hockey stick received prominence in the 2001 IPCC Report. Ross McKitrick wrote:

It was central to the 2001 Third Assessment Report (TAR) from the IPCC. It appears as Fig. 1b in the Working Group 1S Summary for Policy Makers, Fig. 5 in the Technical Summary, Twice in Chapter 2 and in the Synthesis Report, Referring to this figure, the IPCC Summary for Policymakers "that the 1990s has been the warmest decade and 1998 the warmest year of the millennium for the northern hemisphere.

He also notes the importance of the hockey stick to their scientific agenda, designed to support the political agenda, measured by the different highlighting from other information in the report:

In appreciating the promotional aspect of this graph, observe not only the number of times it appears, but its size and colorful prominence every time it is shown.

The control of what went into the Technical Summary (Science Report) and the SPM by just a few people is the real issue and critical to understanding how a few people controlled the deception that fooled the world.

Mann's work provided the handle for the hockey stick. He rewrote history by eliminating the MWP, but the hockey stick has a blade with data provided effectively by Phil Jones, Director of the CRU and IPCC lead author. His work claimed temperatures after the Little Ice Age rose at a rate greater than any in the natural record and thus indicated a human signal. In the SPM the hockey stick and temperature graphs appear together as Fig. 34.

Chapter 20

VARIATIONS OF THE EARTH'S SURFACE TEMPERATURE

Jones claimed an increase of .6 degree C in the global average annual temperature in approximately 130 years. The actual statement in the SPM is odd:

Over both the last 140 years and 100 years, the best estimate is that the global average surface temperature has increase by 0.6 plus or minus 0.2 degrees C.

The IPCC claim this increase is beyond any natural increase and, therefore, anthropogenic.

This is simply incorrect. Actually, it is within the error factor of calculation of global average temperatures. Besides, there are so many problems with the data many consider it impossible to calculate the global temperature. The error range of .2 C, which is plus or minus 33%, shows the problem. The IPCC report identifies some of these.

There are uncertainties in the annual data due to data gaps, random instrumental error and uncertainties, uncertainties in bias corrections in the ocean surface temperature data and also in adjustments for urbanization over the land.

Here are some of the other problems:

There are very few records, approximately 1000, of 100 years length or more, according to the Goddard Institute of Space Studies (GISS).

Most are concentrated in eastern North America and Western Europe. Most of these stations are affected by the Urban Heat Island effect that artificially increases the temperature. Instruments varied regionally over time, but because of early limitations all records only measure to 0.5 degrees. There are virtually no measurements for the oceans, which are 70% of the surface. Few measurements exist for the deserts (19% of the land surface), mountains (20%) or forest (40%).

There is serious scientific concern about the nature, length and quality of the data base best expressed by the US National Research Council Report in 1999.

Deficiencies in the accuracy, quality and continuity of the records place serious limitation on the confidence that can be place in research results.

The actual figures Jones gave, which were 0.6 degrees C plus or minus 0.2 degrees C, an effort factor of 33% underscore this problem. The limitations and the error factor and sufficient to reject the argument that it represents a real increase. It is completely inadequate to serve as part of the basis for global climate and energy problems.

But, there is a more serious problem. We are unable to reproduce Jones results because he refused to disclose which stations he used and how he adjusted the data. On Feb. 21, 2005 in response to a request from Warwick Hughes, an Australian researcher who has long sought to verify the global temperature record, Jones wrote.

"We have 25 years or so years invested in the work. Why should I make the data available to you, when your aim is to try and find something wrong with it?"

Apparently Jones is not alone in the practice of non-disclosure or denial of access to climate data. Finally, we learned from Jones that the original data was lost. This was not the first or the last time, the adjustments to and selection of raw temperature data was "lost".

A series of attempts to obtain information from the University of East Anglia and from the joint enterprise of the Hadley Centre and Climate Research Unit known as the HadCRUT3 appear on the Climate Audit Blog site. This site also discusses disturbing questions about modifications to past records apparently to make the 1930s appear cooler, thus enhancing the claim that the world is warmer than it has ever been.

The most recent 'human signal' is not actual evidence at all. It comes from carefully manipulated computer models designed to isolate a portion of temperature increase as clearly human. People are generally unaware that all predictions of global warming come from computer models. These can't work because the database that limits Jones in totally inadequate for models. Another factor compounds the problem. While Jones estimated surface temperature, the models are three-dimensional and we have virtually no data for the atmosphere. Combine this with the limited knowledge of atmospheric, oceanic, solar mechanisms necessary for a model to work and it is no surprise the models fail to simulate past climates or accurately predict future ones. Models that can't forecast weather beyond 7 days are incapable of predicting conditions 39, 40, or 100 years from now.

It is impossible to address all the errors in the science, assumptions, methods, data, and computer models. So far I have examined the objective, motive and some of the major gaming carried out. However, it is al ultimately tested in the real world. A simple but powerful definition of science is the ability to predict. If your predictions are wrong there is clearly something wrong with your science.

Weather forecast failures indicate it is not a science. Supports of "official" climate science, produced by the IPCC, tried to distance themselves from this problem by saying that they were two different things. The difficulty is climate is an average of the weather; therefore it can only be as precise as the weather.

Every single climate prediction (projection) the IPCC made has been wrong. As we saw earlier, they ostensibly switched from predictions to projections, because of the failures. They then produced three scenarios based upon economic development that would determine the amount of human CO_2 produced. It is likely that much of this was due to manipulation of the major components including the temperature and human production of CO_2.

AR5 Continues to let the End Justify the Unscrupulous Means

Someone said economists try to predict the tide by measuring one wave. The IPCC essentially try to predict (project) the global temperature by measuring one variable. The IPCC compound their problems by projecting the temperature variable with the influence of the economic variable.

Use of circular arguments is standard operating procedure for the IPCC. For Example, they assume a CO_2 increase causes a temperature increase. They then create a model with that assumption and when the model output shows a temperature increase with a CO_2 increase they claim it proves their assumption.

They double down on this by combining an economic model that projects a CO_2 with their climate model projection. To make it look more accurate and reasonable they create scenarios based on their estimate of future developments. It creates what they want, namely that CO_2 will increase and temperature will increase catastrophically unless we shut down fossil fuel based economies very quickly.
All their projections failed, even the lowest as, according to them, atmospheric CO_2 continued to rise and global temperatures declined. As usual, instead of admitting their work and assumptions were wrong, they scramble to blur, obfuscate and counterattack.

———

One part of the obfuscation is to keep the focus on climate science. Most think that the IPCC is purely about climate science, they don't know about the economics connection. They don't know that the IPCC projects CO_2 increase on economic models that presume to know the future. Chances of knowing that are virtually zero, as history shows.

CLIMATEGATE

We now know, through leaked emails from the Climate Research Unit (CRU) at the University of East Anglia, how a small group who were also members of the IPCC, created a totally false picture supposedly based on Science. Some have described what the IPCC achieved as similar to Lysenkoism.

In the Soviet Union, Lysenko guided science not be the most likely theories, backed by appropriately controlled experiments, but by the desired ideology. Science was practiced in the service of the state, or more precisely, in the service of ideology.

Lysenko's version of genetics dominated and seriously diverted Soviet science from 1948 to 1965 until finally rejected. The concept that human CO^2 causes warming and climate change was based on unproven theory used by people with an ideology. They used instruments of state to dominate the science. They also attacked and abused anyone who dared to pursue proper science.

The small group who controlled the IPCC was unlikely to change their tune. A pattern that was borne out by the release of the IPCC report of ARJ5 in 2013 which denied the fact that for 17 years global temperature declined slightly while CO^2 levels continued to increase.

A group, led by Canadian Donna Laframboise, examined the claim by IPPC Chairman Rajendra Pachhauri comment in *The Times of India*:

IPCC studies only peer-review science. Let someone publish the data in a decent credible publication. I am sure IPCC would then accept it; otherwise we can just throw it into the dustbin.

The Laframboise group found that:

21 of 44 chapters in the United Nations' Nobel-winning climate bible earned an F on a report card today. Forty citizen auditors from 12 countries examined 18,500 sources cited in the report-finding 5,600 to be not peer-reviewed.

When trying to identify that people know what was going, it is most effective when it is in their own words. Based on their own comments, most people involved with the IPCC knew from the start the limitations of the data and the climate modes. Kevin Trenberth, when responding to a report on the inadequacies of the weather data produced by the US National Research council, said:

It's very clear we do not have a climate observing system...this may come as a shock to many people who assume that we do know adequately what going on with the climate, but we don't.

A few years ago the Climatic Research Unit (CRU), before it became infamous for leaked emails exposing the manipulation of climate science, issued a statement that said:

General Circulation Models (GCM's) are complex, three

dimensional computer-based models of the atmospheric circulation. Uncertainties in our understanding of climate processes, the natural variability of the climate, and limitations of the GCM's mean that their results are not definite predictions of climate.

Phil Jones, Director of the CRU at the time of the leaked emails, and Tom Wigley, a former director of the CRU and player in the IPCC said:

Many of the uncertainties surrounding the causes of climate change will never be resolved because the necessary data are lacking.

Stephen Schneider, a very prominent part of the IPCC from the start, said:

Uncertainty about feedback mechanisms is one reason why the ultimate goal of climate modelling-forecasting reliably the future of key variables such as temperature and rainfall pattern—is not realizable.

Schneider also set the tone when he said in *Discover* magazine:

"Scientists need to get some broader based support, to capture the public's imagination. That, of course entails getting loads of media coverage. So we offer up scary scenarios, make simplified dramatic statements, and make little mention of any doubts we may have—what the right balance is between being effective and being honest."

I bet you thought scientists are supposed to be honest when they use terms like scientific fact and settled science. Scientists are not supposed to make up "scary scenarios" so that they can be "effective". Scientists who don't tell the truth are politicians masquerading as scientists.

The IPCC achieved Schneider's objective with dramatic but devastating effect. This is not surprising because the first IPCC Chairman, Sir John Houghton said:

Unless we announce disasters no one will listen.

It appears that the IPCC and its disciples decided that the end justifies the means. Phil Jones, Director of the CRU and major player in the IPCC demonstrated this when on July 8, 2004 in response to concerns about papers challenging their "science" he wrote to Michael Mann in a leaked email, with the label "Higher Confidential" that:

I can't see either of these papers being in the next IPCC report, Kevin (Trenberth) and I will keep them out somehow—even if we have to redefine what the peer-reviewed literature is!

They used computer models to produce the results they wanted and bamboozle most people. Even a cursory examination shows they are inadequate, even by the admission of their greatest manipulators, the IPCC.

Perhaps the most insidious activity included controlling climate activity through Wikipedia. Many students and much of the media rely on Wikipedia for research. Most don't know how the material was entered or edited. They are learning partly due to the activities of people connected to the activities of people connected to the Climate Research Unit (CRU).

William Connolley knew and exploited the opportunity. A participant in computer modeling, his activities are shocking. He established himself as an editor at Wikipedia and with a cadre of supporters he controlled all entries relating to climate, climate change and the people involved. This included putting up false material about skeptics. They constantly monitored the entries and rapidly returned to the original false information any attempted corrections. With so many people, they could easily circumvent the limit on the number of edits per person. Connolley as a designated editor had even more latitude. Here is how Lawrence Solomon described the activities.

All told, Connolley created or rewrote 5,248 unique Wikipedia articles. His control over Wikipedia was greater still, however, through the role he obtained at Wikipedia as a website administrator, which allowed him to act with virtual impunity. When Connolley didn't like the subject of a certain article, he removed it. More than 500 articles of various descriptions disappeared at his hand. When he disapproved of the arguments that others were making, over 2,000 Wikipedia contributors who ran afoul of him found themselves blocked from making further contributions. Acolytes whose writing conformed to Connolley's global warming views, in contrast, were rewarded with Wikipedia's blessings. In these ways, Connolley turned Wikipedia into the missionary wing of the global warming movement.

The Medieval Warm Period disappeared, as did criticism of the global warming orthodoxy. With the release of the Climategate emails, the disappearing trick has been exposed. The glorious Medieval Warm Period will remain in the history books, perhaps with an asterisk to describe how a band of zealots once tried to make it disappear.

As the German magazine, *Die Kalte Sonne,* reported in January, 2013:

Unbelievable but true: The Wikipedia umpire on Climate Change was a member of the UK Green Party and openly sympathized with the views of the controversial IPCC. So he was not a referee, but the 12th Man of the IPCC team.

Chapter 21

THE PHYSICAL BASIS OF THE MODELS

The following is how the atmosphere is divided to create models. The surface is covered with a grid and the atmosphere is divided into layers. Computer models vary in the size of the grids and number of layers. The modelers claim a smaller grid provides better results. It doesn't! The model needs data for each cube. However, there are no weather stations for at least 70% of the surface and virtually no data above the surface. There are few records of any length anywhere: the models are built on virtually nothing. In addition. The grid is so large and crude they can't include major weather features like thunderstorms, tornados, or any small storm systems.

They resolve the lack of data using a technique called parameterization. It is simply very poor estimates produced by computer models. For example, they claim using this method that the weather data from a single station is representative of temperatures for a 1200 km radius. Think about using Santa Barbara temperature for Death Valley and San Francisco. Also realize this is an example of how they produce data from one model and then use it as real data in another model.

O'Keefe and Kuester explain how model works:

The climate model is run, using standard numerical modelling techniques, by calculating the changes indicated by the model's equations over a short increment of time—20 minutes in the most advanced GCM's—for one cell, then using the output of that cell as inputs for its neighboring cells. The process is repeated until the change in each cell around the globe has been calculated.

Imagine the number of calculations necessary that even at computer speed of millions of calculations a second takes a long time. The run time is a major limitation. All of this takes huge amount of computer capacity: running a full-scale GCM for a 100 year projection of future climate requires many months of time on the most advanced supercomputer. As a result, very few full-scale GCM projections are made.

A comment at Steve McIntyre's site, Climate Audit, illustrates the problem:

Caspar Amman said that GCM's (General Circulation Models) took about 1 day of machine time to cover 25 years. On this basis, it is obviously impossible to model the Pliocene-Pleistocene transitions (say the last 2 million years using a GCM as this would take about 219 years of computer time.

As a result, very few full-scale GCM projections are made. Modelers have developed a variety of short cut techniques to allow them to generate more results, since the accuracy of full GCM runs is unknown, it is not possible to estimate what impact the use of these short cuts has on the quality of model outputs.

Omission of variables allows short runs, but allows manipulation and removes the model further from reality. Which variables do you include? For the IPCC only those that create the results they want. Also, every time you run the model it provides a different result because the atmosphere is chaotic. They resolve this by doing several runs and then using an average of the outputs.

By leaving out very important components of the climate system, they increase the likelihood of a human signal being the cause of change. As William Kininmonth, meteorologist and former head of Australia's National Climate Centre explains:

Current climate modeling is essentially to answer one questions: how will increase atmospheric concentrations of CO^2 (generated from human activity), change earth's temperature and other climatological statistics? Neither cosmology nor volcanology enters the equations. It should also be noted that observations related to sub-surface ocean circulation, the prime source of internal variability, have only recently commenced on a consistent global scale. The bottom line is that IPCC's view of climate has been through a narrow prism. It is heroic to assume that such a view is sufficient basis on which to predict future climate.

The IPCC take the average results of some 22 computers and

average them to produce an ensemble result. As Robert Brown, physicist at Duke University notes: First—and this is a point that is stunningly ignored—there are a lot of different models out there, all supposedly built on top of physics, and yet no two of them give anywhere near the same results! The title of the article says it all; the ensemble of models is completely meaningless, statistically.

Another problem identified with the models was confirmed in a peer review paper published in *Monthly Weather Review* for July 26, 2013. It finds that: The same global forecast model (one for geopotential height) run on different computer hardware and operating systems produces different results at the output with no other changes.

The problem develops because each model deals with rounding errors differently. These are individually small but because they are numerous the net results are very different outcomes. In the actual studies in the paper they show differences that appear small, until you realize they are for only "10 days' worth of modeling".

Chapter 22

STATIC CLIMATE MODELS IN A VIRTUALLY UNKNOWN DYNAMIC ATMOSPHERE

Knowledge about the atmosphere and lack of data are serious limitations on understanding climate change and building climate models. The atmosphere and lack of data are serious limitations on understanding climate change and building climate models. The atmosphere is three-dimensional and dynamic, so building a computer model that even approximates reality require for more data than exists and much greater understanding of an extremely turbulent and complex system.

Sunlight striking the Earth does not impart the same amount of energy everywhere. First, only one half the globe is receiving sunlight at any given time. Second, the amount of energy input is maximum when the sunlight is vertical to the surface and that is only at one point at any given time.

"We don't have time for a debate with the flat Earth society" President Obama.

It is the climate modelers who use a flat Earth in their models and they are the flat Earth society. This slur implies that skeptics don't believe there is a greenhouse gas effect on temperature. There is broad consensus on this, the skeptics question the sensitivity. Since the climate models use different numbers for the senility, ranging from 1.5 to 5, and then add differing amounts of aerosols to cool the model as desired, it is modelers themselves who are having a debate about the flat Earth that only exists in their computers.

The computer models assume the Earth is a flat sphere with the energy distributed evenly across the surface. The total amount of energy received is the same for the sphere and the disc but the distribution is very different. On the sphere (Earth) much more is received at the Equator, that at the poles—a difference that creates heat at the Equator and cold at the Poles. This is a relatively simple model, but in reality it is enormously complicated by a variety of motions, not least the rotation and tilt, which varies the way the Earth is exposed to sunlight through the year. This change in tilt is given in most text books as 23.5 degrees, but few people know it is constantly changing, thus causing climate change.

Most are unaware that the orbit around the Sun is not a fixed elliptical orbit or that the tilt changes.

Without going into the details of these changes, it is important to know that science has known about these changes for over 100 years, but they are still not in most textbooks or in the public's understanding. It wasn't until the late 1980s that these changes, known collectively as the Milankovitch Cycles, were generally accepted by science.

As the Earth spins through it 24-hour cycle and rotates around the sun annually, the tilt of the earth's axis, relative to the plane of travel around the sun, causes our four seasons (the hemisphere closest to the sun is in summer, while the opposite hemisphere is in winter. The current tilt of 23.4 degrees is in a state of gradual change, shifting between 21.5 and 24.5 degrees every 41,000 years. This is the obliquity cycle. Presently the Earth is moving back to a lesser tilt of 21.5 degrees), which will produce cooler summers and milder winters.

The precession cycle affects the direction of the tilt. Over the course of 22,000 years the axis of the Earth subtly wobbles from side to side. This forces climate change every 11,000 years. The current wobble has placed the southern hemisphere closer to the sun, allowing for slightly warmer summers and cooler, snowier winters. The increased snow and ice has lowered the average annual temperature in Antarctica.

If you go to a planetarium show a popular feature is the effect of precession (wobble) on the Sun Signs of Astrology. If you were born in July, according to astrology, the sun was in Cancer. This was true thousands of years ago when astrology was developed. Now if you were born in July the sun was in Gemini because of precession.

The earth's orbit around the sun, is not a perfect circle, it is an ellipse. The ellipse varies over 100,000 years so that at one point the Earth is a maximum of 94.5 million miles from the sun and at its closest point is 91 million miles. At the farthest distance there is slightly less solar radiation reaching the Earth resulting in a lower average temperature. At its closest the Earth receives slightly more radiation, resulting in more habitable temperatures, as we are experiencing today. Once the Earth passes from its current optimal position, the planet will be in a frigid phase that will stretch thousands of years, perhaps leading to a major ice age. This cycle is called eccentricity.

These three cycles were popularized by Milutin Milankovitch, a Serbian geophysicist who died in 1958. Milankovitch is best known for his theories of ice ages in relation to the three cycles, now referred to as the Milankovitch cycles.

Consider the changes in climate these changes make and how they complicate construction of a model. The IPCC do not include these effects in their computer models because the timescale is not appropriate, but if you are projecting for 100 years, then they are significant relative to the possible fractional impact of human CO^2.

Understanding of the dynamics of the atmosphere is even more recent and incomplete. In 1735, George Hadley used the wind patterns, record by English sailing ships, to create the first upper level diagram of circulation.

HADLEY CELLS

Restricted only to the tropics, these zones became known as Hadley Cells. Sadly, we know little more now that Hadley did. The IPCC illustrates the point in Chapter 8 of the 2007 report:

The spatial resolution of the coupled ocean-atmosphere models used in the IPCC assessment is generally not high enough to resolve tropical cyclones, and especially to simulate their intensity.

The problem for climate science and modelers is that the Earth rotates.

Its rotation around the sun creates the seasons, but the rotation around the axis creates even greater complexity. Without axis rotation, a simple single cell system with heated air rising at the Equator and moving to the poles, then sinking and returning to the Equator, breaks down.

In the 1850s, William Ferrell attempted to improve understanding and proposed a three-cell system that still appears in most textbooks. This model was convenient for teaching, but didn't work when research like tracking nuclear fallout from atmospheric explosions began in the 1960s.

———⁂———

The Ferrell Three Cell model is inaccurate for a variety of reasons, but especially the difference in height of the cells. Few people know the Prop pause, the boundary between the troposphere and the stratosphere is twice as high at the Equator as at the Poles.

VARYING HEIGHT OF TROPOPAUSE IN THE AVERAGE CONDITION

The Tropopause height at the Poles varies between 7 km in winter and 10 km in summer, at the equator the range is 17 to 18 km. The difference in seasonal range is because of the difference in seasonal temperature range, the summer winter difference is greater at the poles. How do you build even those simple dynamics in to a computer model? Remember the heights vary with global temperature and that varies from year to year.

Now we use the Indirect Ferrell Cell, but the most important part is the discontinuity in the Tropopause and Stratospheric-Tropospheric Mixing. This is important, because the IPCC doesn't include the critical connection between the stratosphere and a major mechanism in the upper troposphere in their models.

Due to the computational cost associated with the requirement of a well resolved stratosphere, the models employed for the current assessment do not generally include the Stratospheric-Tropospheric Mixing.

What are the computational costs associated with the requirement of a well-resolved stratosphere? This means they don't know what is going on, but that's true of most of the troposphere.

Climate models are mathematical constructs that divide the atmosphere into cubes. It doesn't matter how many cubes you create for finer resolution because the data is simply not available, especially above the surface. The computer models are incapable of even approximating reality, but wait, there's more.

Divide the atmosphere into two air masses—the dome of cold air over the poles and in between the warm tropical air. The boundary is called the Polar front and marks the latitude of energy balance. In the cold dome more energy escapes that enters creating a deficit. In the warm air more energy enters then leaves creating a surplus. This difference between surplus and deficit drives the atmospheric system. The greatest difference occurs right at the Polar front and as a result the strongest winds are formed, called the jet stream.

A pattern of jet stream waves is named after Carl Rossby. These Rossby Waves change in pattern between zonal or meridinal flow with each creating very different weather patterns. They explain shifts in the viability of weather that has occurred since 2000 as the world cooled. IPCC models failed to understand this, but that is not surprising because it wasn't until 2007 that NASA admitted that the major cause of different ice conditions in the Arctic were due to changing wind patterns. It is just one of a multitude of limitations of their understanding of weather and climate.

Despite this the IPCC are certain about what has and will happen based on computer models that claim to replicate the atmosphere. This is a serious and unjustifiable claim.

Chapter 23

WEATHER AND CLIMATE FORECAST FAILURES

The most damaging evidence of the continued inadequacy of the models was outline in a conference held in May 2008 at the University of Reading, England.

Roger Harrabin, a BBGC reporter with admitted sympathies to the IPCC work, wrote:

I have spent much of the last two decades of my journalistic life warning about the potential dangers of climate change...and reports they...plan a revolution in climate prediction.

What's the revolution? Belief that bigger and faster computers would solve the problems only illustrates their lack of understanding.

Julia Slingo of the University of Reading, a major center for computer climate modelling, who became Chief Scientist for the United Kingdom Meteorological Office said:

We've reached the end of the road of being able to improve models significantly so we can provide the sort of information that policymakers and business require. In terms of computing power it's proving totally inadequate. With climate models we know how to make them much better to provide more information at the local level—we know how to do that, but we don't have the computing power to deliver it.

This is wrong. It doesn't matter how big or fast the computer is if you don't have accurate data or understand climate mechanisms. The cost of computers is a serious limitation to open and effective research. This effectively restricts the work to government agencies, or academics funded by the government, which allow control by bureaucrats with political agendas. It also means it is necessary to stress the planet saving potential of the work. According to Lindzen at MIT, one of his colleagues, lost his government funding after submitting a paper showing no net warming in a 100 years. The letter from the government said his findings were "a danger to the future of humanity".

—∞∞—

There are two devastating reasons why the model output should not be used for anything, especially policy. First, a standard and required test of a model is its ability to predict. Earlier I gave this as the simple definition of science. This is done in models by a process called validation. You design a model by a process called validation. You design a model based on a known period of climate and then have it recreate (predict) another known period.

No IPCC model has been validated—they cannot recreate past climate conditions, therefore, they cannot make valid predictions or even reasonable scenarios. They claim the models are but what they do is keep adding or adjusting variables until the model recreates the situation. This has nothing to do with the actual processes and is called tweaking. The classic example was the attempts to make their models explain the cooling from 1940 to 1980. They did it by adding sulphates, ostensibly from human activities, to act as a cooling agent in the atmosphere. The problem was temperatures started to increase after 1980 but sulfur levels didn't decrease. Second, the most basic assumption about human-caused climate change is that an increase in CO_2 will cause an increase in temperature. Every record shows that temperature increases before CO_2. Despite this, they program computer models so temperature increases if CO_2 increases, but that doesn't deter Harrabin, who concludes by saying:

Political leaders have now agreed that they cannot wait for the modelling uncertainties to be ironed out. They have said they are convinced that emissions should be cut.

This statement underlies the bizarre thinking. The models are the sole source of evidence that there is a problem. They don't work, but we are going to act anyway. It's a standard strategy of environmentalists known as the Precautionary Principle.

How much longer can the IPCC maintain the charade before enough leaders understand the deceptions and shut them down? It is incredible that the IPCC and its manipulation of climate science continue to drive world energy and economic policies.

DISASTROUS POLICIES JUSTIFIED BY FALSE CLAIM
THE CLIMATE MODEL PREDICTIONS ARE CORRECT

IPCC projections are worse than guesswork, which have a chance of being correct. IPCC tried to mask their complete failure by calling them projections and producing a range, but even the low projection is wrong.

Up to about 2000, it was global warming, then when CO^2 continued to increase but when temperatures stopped increasing it became "Climate Change". A 2004 leaked CRU (IPCC) email from the MINNS/Tyndall Centre on the UEA campus said: "In my experience, global warming freezing is already a bit of a public relations problem with the media."

To which Swedish alarmist Bo Kjellen replied: "I agree with Nick that climate change might be a better labeling than global warming".

Proper science would require they consider the null hypothesis that something other than human CO^2 is causing warming. Instead, they moved the goal posts so the IPCC political objective was not abandoned.

IPCC climate models are the basis for claims of accurate predictions. They're inaccurate and the IPCC tell us why in Chapter 8 of the 2007 Physical Science Report. Virtually nobody reads it. Instead they read the deliberately distorted Summary for Policy Maker (SPM) that gives completely unjustified certainty to model projections. The climate predictions are wrong. Using them as the basis for a war on coal, and other draconian energy, economic and environmental policy brings disaster, as most European Countries already know.

Most don't understand models or the mathematics on which they are built; a fact promoters of human caused climate change exploited.

They are used to "prove" to the public that their science is unassailable. They are also the major part of the IPCC work not yet investigated by people who work outside climate science. Whenever outsiders investigate, as with statistic and the hockey stick, the gross and inappropriate misuses are exposed. The Wegman Report investigated the Hockey Stick fiasco but also concluded:

We believe that there has not been a serious investigation to model the underlying process structures not to model the present instrumented temperature record with sophisticated process models.

Of course, it is not necessary to know what is wrong within the models. There failed predictions (projections) are clearly displayed when compared to actual temperature.

Most of the deception appeared in Assessment Report 5 (AR5) released in September of 2013. Despite a stretch of 17 years of slightly declining temperature, which CO_2 levels continued to rise, they increased claims of certainty of their results from 90% to 95%. One claimed the heat was hidden in the deep ocean. Another said 17 years was an inadequate period of measure. The point is the decline could not have happened at all according to their science. Again the IPCC people know because on October 12, 2009, Trenberth said, "the fact is that we can't account for the lack of warming at the moment and it is a travesty that we can't.

Phil Jones, Director of the CRU and deeply involved with what went on there and at the IPCC said:

We don't fully understand how to input things like changes in the oceans, and because we don't fully understand that you could say that natural variability is not working to suppress the warming. We don't know what natural variability is doing.

The entire exercise of global warming and climate change is a deception. However, there are deceptions within the deceptions, with the most important one being that the IPCC produce essentially unassailable scientific predictions using computer models that represent reality.

There is hidden deception cleverly presented to in clear view—a form of daylight robbery. This was achieved primarily by the difference between the definitive statements of the Summary for Policymakers

(SPM) and the Physical Science Basis Report of Working Group I. The comments about the limitations of the models in Chapter 8 are in stark contradiction to the claims in the SPM.

They underscore the political nature of the deception. They know very few would read or understand the science report and they could be easily marginalized. With limitations in the science report, they could, if challenged, say they warned everybody.

When they compared the SPM to the AR4 Synthesis Report Armstrong and Green determined it:

Is a political document that downplays assessments of uncertainty from the scientific reports written by the IPCC, which themselves are far more subjective than the IPCC would have one believe. Equally important, both the IPCC's summaries and main reports omit much contrary evidence.

It is only one of several deceptions, but unquestionably the worst because it was so effective.

Computer models produce IPCC projections as a central part of every Report. The process is progressive with each Report building on the last. They also, supposedly, introduce new findings. The difficult is inclusions are selective because of the narrow definition of climate change and the rules that set a last date of publication. There are instances of selective inclusions beyond the deadline or exclusion when the material was available in time.

The IPCC produced the SPM for policymakers. Cleverly and cynically they disclosed the limitations of their work in the Physical Science Report of Working Group I. This means that nobody can say they didn't acknowledge the limitations. One of the most serious is the lack of data and the way it is created. In the 2007 IPCC AR4, Chapter 8 Advances in Modelling they say:

Despite the many improvements, numerous issues remain; many of the important processes that determine a model's response to changes in radiative forcing are not resolved by the model's grid. Instead, sub-grid scales are used to parameterize the unresolved processes, such as cloud formation and mixing due to oceanic eddies. It continues to the case that multi-model ensemble simulations generally provide more robust information than runs any single model.

Parameterization is a fancy word for making up data when it doesn't exist. Because of the lack of data, this occurs for a majority of the surface grids and virtually all of the layers above the surface. This means they estimate average value for each grid and use those as the base for the model. Computer models produce estimates, which they use as "real' data in another. Parameterization is probably applied to 80% of the surface and at least 90% in the atmosphere. Even with the grid scale used it is too large for major weather event such as thunderstorms.

The climate system includes a variety of physical processes, such as cloud processes, radiative processes and boundary layer processes, which interact with each other on many temporal and spatial scales. Due to the limited resolutions of the models, many of these processes are not resolved adequately by the models grids and must therefore by parameterized. The differences between parameterizations are an important result why climate model results differ.

The problem is bigger than just cloud formation. For example, thunderstorms in the tropics are too small for the grids but they are a major mechanism of balancing and cooling the atmosphere. They distribute surplus heat from the tropics to offset deficit heat energy in the Polar Regions.

There is currently no consensus on the optimal way to divide computer resources among finer numerical grids, which allow for better simulations; greater numbers of ensemble members, which allow for better statistical estimates of uncertainty' and inclusion of a more complete set of processes (e.g. carbon feedbacks, atmospheric chemistry interactions. They are saying there is inadequate computer capacity and large amounts of data are excluded and major mechanisms ignored.

As A.N. Whitehead said:

"There is no more common error than to assume that because prolonged and accurate mathematical calculations have been made, the applications of the result to some fact of nature is absolutely certain."

At a conference in Edmonton on Prairie Climate Predictions, climate modeler, Michael Schlesinger, dominated as the keynote speaker. His presentation compared five major global models and their results. He claimed that because they all showed warming they were valid. Of course they did because they were programmed to that general result. They problem is they varied enormously over vast regions. For example, one showed North America cooling, while another showed warming. The audience was looking for information adequate for planning and became agitated, especially in the questions period. It peaked when someone asked about the accuracy of his warmer and drier prediction for Alberta. The answer was 60%. The person replied that is useless, my Minister needs 95%. The shouting intensified. Eventually a man threw his shoe on the stage. When the room went silent he said, "I didn't have a towel". He asked permission to go on stage where he explained his qualifications and put a formula on the blackboard. He asked Schlesinger if this was the formula he used as a basis for his model of the atmosphere. Schlesinger said yes. The man then proceeded to eliminate variables asking Schlesinger if they were omitted in his work. After a few eliminations he said one was probably enough, but you have no formula left and you certainly don't have a model. It has been that way ever since with the computer models.

The IPCC only include those that create the results they want, namely proof of human causes of Climate Change. Also, every time you run the model it provides a different result because the atmosphere is chaotic. They resolve this be doing several runs and then using an average of the outputs.

As noted earlier, by leaving out very important components of the climate system they guarantee a human signal will result. As William Kininmonth, meteorologist and former head of Australia's National Climate Centre describes them:

"…current climate modeling is essentially to answer one question: how will increase atmospheric concentrations of CO^2 (generated from human activity) change earth's temperature and other climatological statistics? Neither cosmology nor volcanology enters the equations. It should also be noted that observations related to sub-surface ocean circulation (oceanology), the prime source of internal variability, have only recently commenced of a consistent global scale. The bottom line is that IPCC's view of climate has been through a narrow prism. It is heroic to assume that such a view is sufficient basis on which to predict future 'climate'".

"Heroic" is an understatement. When you leave variables out of an equation, you no longer have an equation because one-side cannot equal the other. For example, the IPCC 2007 Report on their computer models of the atmosphere says, "Due to the computational cost associated with the requirement of a well-resolved stratosphere, the models employed for the current assessment do no generally include the QBO". This is the Quasi-Biennial Oscillation of upper level winds related to the pattern of El Niño.

Chapter 24

THE MYTH OF
MORE POWERFUL COMPUTERS

Billions of taxpayer's dollars are wasted on politically motivated climate science, most of it on buying and running computers incapable of modeling global climate. They are incorrectly programmed so a CO^2 increase causes a temperature increase to fulfill the saying Garbage In-Garbage Out. In the real world temperature increase before CO^2, but the programmers need a political result. Naturally the temperature forecasts are consistently wrong, but that doesn't matter. The claim they're getting better and all they need are bigger, faster computers. It won't and can't make a difference, but they continue to waste money.

Recently Cray computes produced the Gaea supercomputer for climate research at the National Oceanic Atmospheric Administration (NOAA). More commonly spelt Gaia, after the Greek Earth goddess of the religion of environmentalism. It will produce meaningless results like all the computers despite being on the biggest and fastest. It has a 1.1 petaflops capacity. FLOPS means Floating-Point Operations per Second and peta is a thousand million for 1.1 thousand million floating point operations a second. This sounds impressive, but is totally inadequate. At the Reading conference Shukla reported:

The current generation high-end computers for climate research have a capability of about 50 teraflops, which makes it possible to integrate a typical climate model with about 100 km horizontal resolution for 20 years in one day.

So at 50 trillion floating point calculations per second they only study 20 years of record per day. Worse, each run using identical input yields different results so they average several runs. This is with a grossly simplified model with a grid so large that each covers very different climate regions. Shukla challenges: We must be able to run climate models at the same resolution as weather prediction models, which may have resolutions of 3-5 km within the next 5 years. This will require computers with peak capability of about 100 petaflops.

It makes no difference: weather prediction models don't work either. Proponents argue that weather predictions are different than climate predictions. They're not because climate is the average weather; wrong weather yields wrong climate.

Maurice Strong set up the IPCC through the World Meteorological Organization (WMO), which provided access to national funding for expensive machinery. It also meant appointment of climate modelers often called scientists; their work has little in common with traditional science.

Nor has the IPCC subjected climate models to rigorous evaluation by neutral, disinterested parties. Instead it recruits the same people who work with these models on a daily basis to write the section of the Climate Bible that passes judgment on them. This is like asking parents to rate their own children's attractiveness.

The relationship between one country's climate models and the IPCC illustrates this point. George Boer is considered the architect of Canada's climate modeling efforts. As an employee of Environment Canada, he has spent much of this career attempting to convince the Powers that be that climate models are a legitimate use of public money.

They are not. Canada's Auditor General identifies 6.36 billion climate change funding announcements between 1997 and 2005, but at what price? A December 13, 2011 story provides an answer. Environment and Sustainable Development Commissioner Scott Vaughn reports:

Environment Canada has failed to implement a strategic plan to improve its internal scientific research in areas ranging from managing air and water pollution to toxic chemicals.

Billions are spent on useless computers and climate change while not dealing with real problems. They're not alone, it's happening in national weather agencies round the world.

Despite all that money spent by Environment Canada, they have the worst performing computer climate models, hotter than the IPCC ensemble average, which is hotter than the actual temperature.

The difference between what the IPCC acknowledges is missing or inadequate in order to construct Climate Models and the certainty of their conclusions is so stark it suggests that it is clearly premeditated. Here a few limitations that the IPCC manipulated to cover inadequacies or create a result.

There are major limitations of data like temperature, as Watts and D'Alco explain in their detailed study of global records. Knowledge of precipitation is even worse. When writing about precipitation data Bob Tisdale says, "Amazingly, there are few to no agreements among the three datasets when looking at Global Data."

In 2006 an attempt was made to forecast summer monsoon rains for the Sahel in Africa. They failed and concluded:

Climate scientists cannot say what has delayed the monsoon this year or whether the delay is part of a larger trend. Nor do they fully understand the mechanisms that govern rainfall over the Sahel.

Worse, when they projected long-term trends one model said wetter, the other drier. Why?

One obvious problem is a lack of data. Africa's network of 1152 weather watch stations, which provide real-time data and supply international climate archives, is just one-eighth the minimum recommended by the World Meteorological Organization (WMO). Furthermore, the stations that do exist often fail to report.

They also concluded that the resolution of the models was inadequate and unaware of the underlying mechanisms.

However, there are many lesser variables critical to understanding what creates and determines the weather. Analogies are useful, but dangerous: witness the analogy of the atmosphere to a greenhouse.

One limitation is the role of moisture in transferring heat and energy not considered in the greenhouse yet critical in the atmosphere.

Consider the role of moisture in controlling human body temperature. Few people know the skin is the largest organ of the body, performing an important function as the interface between the inner body and the atmosphere and controlling body temperature.

The amount of moisture used in sweating is a fraction of the total amount in the body, but critical to controlling temperature.

The land and water surface of the Earth are the interface or skin between the subsurface and the atmosphere. The Earth is heated by shortwave energy from the sun that is stored in the surface. Energy for sweating is taken from the body, which cools the body. Evaporation of water from the surface takes heat stored in the Earth or water and transfers it to the atmosphere. The amount of soil moisture is estimated as "0.001% of the total water found on Earth, but like the sweat, it is critical to controlling temperature.

The argument is belied by comments about missing variables such as this one from NASA that confirms the importance of soil moisture:

Soil moisture is a key variable in controlling the exchange of water and heat energy between the land surface and the atmosphere through evaporation and plant transpiration. As a result, soil moisture plays and important role in the development of weather patterns and the production of precipitation.

In 1992 the National Research Council suggested why weather forecast models don't work. Despite the importance of soil moisture information, widespread and/or continuous measurement of soil moisture is all but nonexistent.

The lack of a convincing approach of global measurement of soil moisture is a serious problem.

This was in 1992 but, little had changed when Chapter 8 of the 2007 IPCC Report appeared.

Since the TAR, there have been few assessments of the capacity of climate models to simulate observed soil moisture. Despite the tremendous effort to collect and homogenize soil moisture measurements at global scales (Robock et al, 2000), discrepancies between large-scale estimate of observed soil moisture remain.

Global climate models are composites of individual modes for each component of the atmosphere and assume the smaller models output are real data and they know how it interacts with all other model inputs.

Even that is a problem as Koster, et al, explain:

The soil moisture state simulated by a land surface model is a highly model-dependent quantity, meaning that the direct transfer of one model's soil moisture into another can lead to a fundamental, and potentially detrimental, inconsistency.

The NASA statement identifies another limitation of the IPCC model when they refer to "the exchange of water and heat energy between the land surface and the atmosphere".

Again the IPCC model fails as Chapter 8 notes:

Unfortunately, the total surface heat and water fluxes are not well observed.

This means they don't have the data, but they also admit they cannot simulate the mechanisms involved.

For models to simulate accurately the seasonally varying pattern of precipitation, they must correctly simulate a number of processes (evapotranspiration, condensation, transport) that are difficult to evaluate at a global scale.

Limitations of soil moisture data and mechanisms in the computer models invalidate any output they produce, but it's only one of many. What is outrageous is these gross inadequacies do not stop them claiming that:

Most of the observed increase in globally averaged temperatures since the mid-20th century is very likely due to the observed increase in anthropogenic greenhouse gas concentrations.

Most and very likely are greater than 90% by their definition. These are high levels of certainty even in research based on solid data with reasonable understanding of the mechanisms. They're totally unjustified from the computer model inputs and outputs and the failure of every single prediction or scenario. Magnitude of the disparity suggests those who produce it are either scientifically incompetent or have created the result.

A large portion of the solar heat at the heat Equator is used for evaporation, changing the water from liquid to gas (water vapor). The heat used isn't lost but stored as latent heat and transported on the wind systems. Transfer of energy between the surface and the atmosphere, known as flux, is a major problem in the IPCC models. Their 2007

Report notes:

Unfortunately, the total surface heat and water fluxes are not well observed.

Translation: they don't know how much heat and water moves in and out of the earth's surface. They acknowledge it creates another problem.

These errors in oceanic heat uptake will also have a large impact on the reliability of the sea level rise projections.

But threats of sea level rise are a major part of the fear and exploitation of the IPCC and their fellow Nobel winner Al Gore. Al Gore has predicted a 24 ft. sea rise in 100 years. The IPCC predicts 24 inches. This is known in science as a unit's problem.

As the air rises it cools and condenses. The water vapor converts back to liquid and the latent heat is released into the atmosphere. In the tropics this creates the major cloud form of cumulonimbus (thunderstorms), massive towering structures with powerful internal winds carrying vast amounts of energy through the atmosphere toward the Poles.

Global climate models divide the world surface into large rectangles. Essex and McKitrick note: Not only can we not handle today's thunderstorms, but no such storm ever shows up, even in our very best computer climate models. Thus thunderstorms certainly are not dealt with from first principles in climate models either.

The difficulty is:

At every moment, there are thousands of active thunderstorms in the hot, moist places of the planet.

There are tens of millions of them in any year. It should be clear that this great and constant roar of atmospheric air-conditioning is an important part of the global energy budget and should figure significantly into any model of the global climate. However, the mighty creature overhead, along with all his cousins, is too small to show up in even the biggest and grandest global climate model.

Essex and McKitrick comment:

People who do serious climate calculations understand this problem and the fundamental scientific dilemma it implies. The only way to produce non-absurd calculations is to make up some ad hoc rules that insert or take away the energy, moisture or momentum has needed to produce sensible behavior.

Even so, these made-up rules are not foolishly done. From the collective effects of sub-grid scale phenomena, parameterizations, empirical rules that mimic the overall effect of these phenomena fairly closely—are introduced.

But they can't be accurate because the basic data is lacking and mechanisms inadequately understood. As the IPCC report not each model is parameterized differently so:

The differences between parameterizations are an important reason why climate model results differ.

Inadequacies of modeling the Hadley Cell, a major mechanism introducing weather and climate, are enough to invalidate the models and cause the failed predictions. The IPCC claim that:

Most of the observed increase in global average temperatures since the mid-twentieth century is very likely due to the observed increase in anthropogenic GHG concentrations.

In IPCC jargon "very likely" means more than 90% certain, but inadequate modeling of the Hadley Cell alone makes that an utterly false claim. When the models projections failed, as they did spectacularly in the 2013 Report, they only increased their certainty.

TEMPERATURE DATA AND DATA MANIPULATION

Hubert Lamb, who many consider the father of modern climate studies, established the Climate Research Unit (CRU,) in his autobiography Through All the Changing Scenes of Life he said that he created the CRU because:

It was clear that the first and greatest need was to establish facts of the past record of the natural climate in times before any side effects of human activates could well be important.

The amount of data is still inadequate, which poses a problem far beyond Lamb's concern.

By-predetermining the outcome of their scientific investigation the IPCC were increasingly forced to ignore, counteract, misinterpret, misinform and eventually adjust the data. It became increasingly necessary to become more devious. We have already discussed how their actions completely contradicted the scientific method of trying to disprove hypotheses. The search was for data and research that proved the hypothesis even if it had to be manufactured. To illustrate the point, consider that in every case when they adjusted temperature data it always made the past colder and the present warmer.

The hockey stick was research designed to prove that the 20th century was the warmest in history. In that case, they eliminated the well-established Medieval Warm Period (MWP). They rewrote history to prove their hypothesis. It is not the only example, albeit the most egregious. A global program to lower temperatures in the early part of the instrumental record changed the slope of the temperature curve. This made temperatures at the end of the 20th century appear warmer and suggested a more dramatic increase.

The Earth has generally warmed since the 1680s. Politics has made that period, which covers the Industrial Revolution, of climatic interest as people wanted to prove that human production CO^2 by their industry was causing warming. Proponents ignore the natural warming since the nadir of the Little Ice Age in the 1680s.

Manipulation of records occurs by a variety of means including:
  -Filling in missing data incorrectly or inappropriately
  -Selecting stations that provide the desired result
  -Incorrectly adjusting for instrument changes used over time
  -Failing to or incorrectly adjusting for artificial heating by the urban heat island effect (UHIE)

The latter is an increase in temperature at the weather station as urban development surrounds the station. Professor Richard Muller tried to settle the problems in what also turned out be deceptive with the Berkeley Earth Surface Temperate (BEST) project. It said:

Our aim is to resolve current criticism of the former temperature analysis, and to prepare and open record that will allow rapid response to further criticism or suggestions.

Their actions and results altogether belie this claim and point to a political motive. The entire handling of their work has been a disaster. It is not possible to say it was planned, but it thoroughly distorted the stated purpose and results of their work. The actions are almost too naïve to believe they were accidental, especially considering the people involved in the process. Releasing reports to mainstream media before all studies and reports are complete is unconscionable from a scientific perspective. They replicate the deceptive practices IPCC uses of releasing the Summary for Policy Makers (SPM) before the Scientific Basis Reports with which it differs considerably.

Like the IPCC, the BEST panel appears deliberately selected to achieve a result or least ensures a bias. It begins with the leader Richard Muller, who historically supported the AGW hypothesis. There is only one climatologist in the group, Judith Curry, who only recently shifted from a vigorous pro AGW position to a more central and conciliatory position indicating awareness of the political implications. Involvement in the BEST debacle and especially her admission that early release of some results, before the peer-reviewed articles and supporting documentation, in other words the IPCC approach, was at her suggestion is troubling. Ms. Curry's comments indicate a very peripheral involvement in the entire process. This appears to suggest her participation was for public relations and supported by her comment in the data portion of the work. "I have not had hands-on the data". Failure to include a skeptical climatologist appeared to confirm the political objective.

Climate is average of the weather, so it is inherently statistical. Climatology changed from simple analysis of average conditions at a location or region to a growing interest in the change over time.

The instrumental record is so replete with limitations, errors and manipulations that it is not even a crude estimate of the pattern of weather and its changes over time. Failure of all predictions, forecasts, or scenarios from computer models built on that database, confirm the inadequacies.

Problems start with the assumption that the instrumental

measures of global temperature can produce any meaningful results. They cannot. Coverage is totally inadequate in space and time to produce even a reasonable sample.

There are few stations in Antarctica, most of the Arctic Basin, the deserts, the rain forests, the boreal forest and the mountains. Of course, none of these equal the paucity over the oceans that cover 70% of the world. It's a bigger problem in the Southern Hemisphere, which is 80% water.

There was a great reduction in stations after 1960. BEST intended to offset Ross McKitrick's evidence that much of the warming in the 1990s was due to a reduction in the number of stations. As the number of stations fell the recorded warming accelerated. Many stations were removed because they showed cooling, for example removing a mountain station from La Paz, Peru caused a huge jump in average warming for the area.

Temperatures, sometimes to four decimal places, are used as if they are real, measured numbers. Temperatures were recorded to half a degree because until thermocouple thermometers appeared any greater precision was impossible.

Most of the land data is concentrated in Western Europe and eastern North America so these latitudes dramatically over-represent the record. This is noteworthy because climate change is reflected most in these latitudes as the Circumpolar Vortex shifts between Zonal and Meridional flow and the amplitude of the Rosby Waves vary.

BEST used a subset of global temperatures, albeit a larger subset than anyone else. Because the full data set is inadequate, a bigger subset does not improve the analysis potential. Also, those who used smaller subsets did so to create a result to support a hypothesis. The BEST study apparently was designed to confirm the results and negate criticisms.

Regardless of the BEST findings, the other 3 agencies achieved different results by the stations they chose, and the differences are significant. For example, one year there was a difference of 0.4 degree C between their global averages, which doesn't sound like much, but consider this against the claim that a 0.7 degree increase in temperature over the last approximately 130 years. What people generally ignore is that in the IPCC estimate of global temperature increase produced by Phil Jones of 0.6 degree C, the error factor was plus or minus 0.2 degree C. That is a plus or minus 33% error rate, which makes the record and results meaningless because in many years the differences in global annual average temperature determined by different agencies is at least half the 0.7 degree figure. In summations, all 4 groups selected subsets, but even if they had used the entire data set they could not have achieved meaningful or significant results.

The use of the phrase "raw temperature data" is misleading. What all groups mean by the phrase is the data provided to a central agency by individual nations. Under the auspices of the World Meteorological Organization (WMO), each nation is responsible for stabling and maintaining weather stations of different categories. The data these station record is the true raw data. However, individual national agencies adjust the data before submission to the central record. They didn't use all stations or all data from each station. However, it appears there were some limitations of the data that they didn't consider, as the following quote indicates. Here is a comment in the preface to the Canadian climate normal 1951 to 1980 published by Environment Canada:

No hourly data exists in the digital archive before 1953; the averages appearing in this volume have been derived from all available "hourly" observations, at the selected hours, for the period 1953 to 1980, inclusive. The reader should note that many stations have fewer than the 28 years of record in the complete averaging.

BEST adjusted the data, but they are only as valid as the original data. For example, the National Institute of Water and Atmospheric Research adjusted the raw data lowering the early temps and thus showing more warming.

The same "adjustments were done in Maastricht. It appears that BEST began with a mindset to find warming and it seriously tainted their work. BEST said:

Berkeley Earth Surface Temperature aims to contribute to a clearer understanding of global warming based on a more extensive and requires analysis of available historical data.

The terminology indicates prejudgment. Why global warming? It doesn't even accommodate the shift to climate change. Why not just refer to temperature trends?

The project indicates a lack of knowledge or understanding of inadequacies of the data set in space or time or subsequent changes and adjustments. Lamb spoke to the problem when he established the Climatic Research Unit (CRU). BEST confirms Lamb's concerns. The failure to understand the complete inadequacy of the existing temperature record is troubling. It appears to confirm that there is incompetence or a political motive, or both.

Consider the case of the United States Historical Climate Network (USHCN) and its history of changing data identified by Steve Goddard. As he notes:

"USHCN2 uses a three step process to cool the past and warm the present. Going from the actual measured daily data to 'raw monthly' reduces the decline. The time of observation adjustment flips the trend from cooling to warming, and then a final mysterious adjustment creates a strong warming trend." On his website he shows the raw data with falling temperature and final temperature chart which shows warming.

Anthony Watts joined with meteorologist Joseph D'Aleo to produce a comprehensive analysis of global temperatures, titled *Surface Temperature Records; Policy Driven Deception?* It is a devastating critique of the inadequacies of surface temperature record yet it is the basis of the entire IPCC studies and models.

SUMMARY OF THE
DISTORTED TEMPERATURE RECORD

1. Instrumental temperature data for the pre-satellite era (1850-1980) have been so widely, systematically, and unidirectional tampered with that is cannot be credibly asserted there has been any significant "global warming" in the 20th century.
2. All terrestrial surface-temperature databases exhibit truly serious problems that render them useless for determining accurate long-term temperature trend.
3. All of the problems have skewed the data so as to substantially overstate observed warming both regionally and globally.

4. Global terrestrial temperature data are gravely compromised because more than three-quarters of the 6,000 stations that once existed are no longer reporting.
5. There has been a severe bias towards removing higher-altitude, higher-latitude, and rural stations, leading to a further serious overstatement of warming.
6. Contamination by urbanization, changes in land use, improper siting, and inadequately calibrated instrument further overstates warming.
7. Numerous peer-reviewed papers in recent years have shown the overstatement of observed longer term warming is 30-50% from heat-island contamination alone.
8. Cherry picking of observing sites combined with interpolation to vacant data grids may make heat-island bias greater than 50% of 20th century warming.
9. In the oceans, data are missing and uncertainties are substantial. Comprehensive coverage has only been available since 2003 and shows no warming.
10. Satellite temperature monitoring has provided an alternative to terrestrial station in compiling the global lover-troposphere temperature record. Their findings are increasingly diverging form the station based constructions in a manner consistent with evidence of a warm bias in the surface temperature record.
11. NOAAA and NASA, along with CRU, were the driving forces behind the systematic hyping of the 20th century "global warming".
12. Changes altered the historical record to mask cyclical changes that could be readily explained by natural factors like Multidecadal Ocean and solar changes.
13. An inclusive external assessment is essential of the surface temperature record of CRU, GIOSS and NCDC, "chaired and paneled by mutually agreed to climate scientist who do not have a vested interest in the outcome of the evaluations."
14. Reliance on the global data by both the UNIPCC and the US GCRP/CCSP requires a full investigation and audit.

Orland, CA—Well Located Station
MMTS Shelter (Maximum/Minimum Temperature System)

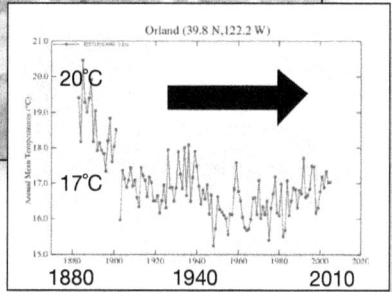

Weather Station

Orland (39.8 N,122.2 W)

20°C

17°C

1880 1940 2010

Marysville, CA—50 miles away

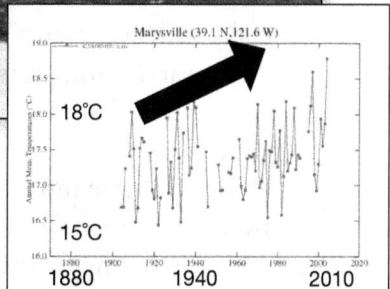

Air Conditioning Units
Exhaust Fans

Weather Station

Asphalt

Marysville (39.1 N,121.6 W)

18°C

15°C

1880 1940 2010

Chapter 25

THE MALTHUSIANS HURT THE POOR IN ORDER TO SAVE THE PLANET

At a 2004 conference of the Russian National Academy of Sciences, Sir David King, and Chief Scientific Adviser to Tony Blair's government made the startling statement that global warming is worse than terrorism. He was right, but not as he intended. The false premise promoted by the IPCC that human CO_2 caused global warming was used to terrorize and undermine developed nations.

The former President of the Czech Republic, Vaclav Klaus, was the only world leader to understand the science and speak out about what Malthusians like Maurice Strong and his IPCC were doing. He was also immediately aware of communism and recognized what was happening. In a 2008 article for *The Australian,* he wrote:

I am afraid there are people who want to limit economic growth, the rise in the standard of living (though not their own) the ability of man to use the expanding wealth, science and technology for solving the actual pressing problems of mankind, especially of the developing countries. The ambition goes very much against past human experience which has always been connected with a strong motivation to better human conditions. There is no reason to make the change just now, especially with arguments based on such in complete and faulty science.

For years Malthusians wearing the green mask of environmentalism have lectured us that the United States has an insatiable lust for oil. And their mantra, constantly repeated by politicians is not one more barrel of oil. We consume more oil than any other country, which is used on our oversized cars, monster trucks and luxurious airplanes. Often the lecturers are rich hypocrites, with multiple airplanes, cars and houses.

Harrison Ford has five airplanes and shaved his chest in a commercial urging people to fight Climate Change. Al Gore often travels with several black Cadillac's and then changes into a Prius several blocks from his public appearance. Al Gore flew to an economic conference in Davos in his private jet in 2015 and announced a campaign to cut carbon emissions.

Our fossil fuel consumption has made us the world's most productive economy and we have done a remarkable job cleaning the air.

Statistics compiled by the American Enterprise Institute reveal that in the 25 years spanning 1980 and 2005:

1. Fine particulate matter declined 40%.
2. Ozone levels declined 20% and days per year exceeding the 8 hour ozone standard fell 79%.
3. Nitrogen dioxide levels decreased 37%, sulfur dioxide dropped 63% and carbon monoxide concentrations were reduced by 74%.
4. Lead levels were reduced by 96%.

What makes theses air quality improvements even more noteworthy is that they occurred during a period which:

1. Automobile miles driven each year nearly doubled to 93% and diesel truck miles more than doubled to 112%.
2. Tons of coal burned for electricity production increased over 60%.
3. The real dollar value of goods and services (GDP) more than doubled to 114%.

Despite more people and more cars on the road than ever, air pollution is no longer a growing problem in the United States. While the cleanup of our skies has gone under reported, the Malthusians now claim, that the air is polluted with CO^2. The American Lung Association runs TV ads claiming CO^2 causes lung disease which is a blatant falsehood. CO^2 causes no diseases; the specification for CO^2 in buildings is three times the current level of 400 ppm.

There are potentially harmful pollutants, not CO^2, associated with coal, but since the Sixties, through technological advances applied in the United States, dangerous impurities such as sulfur and nitrogen oxides and particulates (soot) have been reduced by 90%. Regardless of this verifiable coal clean-up, the Energy Secretary, Steven Chu, has repeatedly said, "Coal is my worst nightmare". Fully aware that soot, sulfur and nitrogen oxides are no longer associated with coal in the United States, his focus is on CO^2.

Because of the falling price of natural gas, the percentage of electricity generated from coal has fall from over 50% to about 38% in 2014. This has resulted in falling CO^2 emissions in the United States. Rather than let market forces continue to work, the EPA is imposing carbon capture technology on coal plants which does not exist in economic form. Thus the real goal of the EPA is to shut down coal plants. Candidate Obama was very candid. He said you can buy a coal plant, but we are going to make it so expensive to operate you will have to shut it down during the 2008 campaign. In the 2012 campaign, when he was in coal country like Kentucky, he said he was for coal. Once elected, he sent the EPA after the coal industry in a war on coal.

During the 2008 campaign, candidate Obama told the New Hampshire editorial board, "I don't think there is anything we inevitably dislike about nuclear power. We just dislike the fact that it might blow up, and irradiate us, and kill us." Once elected, President Obama said his energy policy was "all of the above", meaning wind, solar, natural gas, oil, coal, nuclear etc. Then his first budget proposal cut off all funding for the nuclear waste repository. This repository began construction in Yucca Mountain, Nevada in 1983; in 2014 with 14 billion invested, it was pronounced ready for use. It is still blocked in 2015.

Currently, all nuclear waste created by all of America's nuclear plants since the 1960's, are stored at 126 different sites scattered about the nation. The total amount of spent uranium, the primary waste material amounts to 57,000 tons. Extremely dense, a chunk of uranium the size of a gallon of milk weighs 150 pounds, the 13 gallon jugs of uranium would equal a ton. Thus the 57,000 tons of waste would fit into 760,000 milk jugs, which could fit within the confines of the average high school basketball gym.

So even if we never build another nuclear plant, the waste should be safely stored in Yucca Mountain. For the Malthusians, the nuclear and monetary waste is not the issue—inexpensive, plentiful power is. Paul Ehrlich summed it up saying, "Giving society cheap, abundant energy—would be the equivalent of giving an idiot child a machine gun". Who are the idiot children? In the Malthusian Nirvana, elites like Stanford professors and politicians, would have to travel to world conferences to plan how to protect the "idiot children" from rising out of poverty via cheap and abundant energy.

Reports of the IPCC, falsely presented as based on the science, were used to scare the world, initially about global warming and later climate change. Politicians caught up with the need to appear green grasped at the output of the IPCC. They were thus vulnerable and easily fooled because they didn't understand and the entire objective of the IPCC was to mislead, misdirect and distort.

———⊶⊷———

Instead of helping poor countries and poor people the machinations of the IPCC are helping crony capitalists like Al Gore reap the rewards of their activities while the people pay the price.

Al Gore said he left the White House with fewer than two million dollars of net worth. His net worth is currently more than 100 million dollars. He is one of the biggest carbon capitalist's, owning much of the carbon credit industry.

In California, many people were surprised to find the smart meters were going to be installed and replace meter readers. They were surprised to learn that this was financed by the stimulus program, the 800 million dollar program that was financed with borrowed money. Few people knew that Gore had invested in a smart meter company and then used his political connections to get smart meters inserted into the stimulus funding. Even fewer considered the irony of a stimulus program which caused the loss of many meter reader jobs.

Smart meters are the enabling technology to allow central planners to control energy consumption. Smart meters can be used for peak use pricing. Peak electrical use is often at 3 p.m. on a hot day, because of use of air conditioners. Central planners can apply draconian price increases in order to force reduction of energy use. To a Malthusian, if you let people use all the energy they want, you are rushing to the limits of natural resources, destruction and death. The fact is that this scenario has never come true in 216 years, ironically, there are more Malthusians now than ever before, and they wear the green cloak of environmentalists.

Governments, like California, use carbon taxes and subsidies (called carbon trading) and many restrictive, punitive and expensive regulations to raise huge sums of money. Governor Brown has recently grabbed millions of dollars and diverted carbon trading dollars to fund the high speed rail project.

A good example of the carbon trading program is the Tesla electric car. For every $70,000 car Tesla produces they receive $30,000 dollars in carbon subsidies. The irony of poor people paying carbon taxes every time they buy gasoline, in order to subsidize rich people's purchase of an expensive car, is lost on the Malthusian's wearing the green mask.

The UK newspaper, *The Telegraph* reported:

Investing in climate change is proving to be profitable for governments, corporations, and investors from many sectors. Government's recent subsidies towards energy-efficient programs are bringing in newfound wealth for investors.

Global warming provided the perfect vehicle for Malthusians to spread their claim of human destruction of the planet. Previously they could only point at local or regional problems, but now they had a genuine "the sky is falling" cause that encompassed the entire globe. The demand was for global policies and Maurice Strong led the charge at the Rio Conference in 1992 in the formation of the United Nations Framework convention on Climate Change.

This agency created the Kyoto Protocol that became the battleground. Only the industrialized countries where Strong sought to limit growth had to reduce CO^2 emissions. The excluded Developing nations and arranged payments from the sinful industrialized nation as penance. The futility of the exercise was that if all nations participated and met their original targets no measurable difference in atmospheric CO^2 would occur yet that was the purported objective. Several nations saw the problems implementing Kyoto would create. The US was willing to sign, if it got credit for its vast forests which act as carbon sinks. The Europeans and Strong would not allow this, since it would not limit growth. The US Senate voted 95-0 against ratification even though Al Gore was Vice President at the time. It reached a critical point when a failure of Russia to sign meant the Protocol would not be implemented.

———⊗⊗⊗———

The protocol is now dead and there is no world climate treaty. The countries excluded from the Protocol, particularly India and China, have become the dreaded industrialized nations that the Malthusians oppose. The naiveté and political tunnel vision of the Malthusians ignored the fact that every country in the world wanted to industrialize and emulate the United States.

In 2005, the *Pittsburgh Tribune* reported:

Recently, Maurice Strong (founder of the IPCC) was looking for an apartment in Beijing, where his Canadian interests are already enmeshed with the Chinese Red Army.

A 2006 report said he formed a company with George Soros to import Chinese made cars into the North American market. At the *Tribune* summarized:

Maurice Strong is the fox that was invited into the henhouse—and given the tools to redesign it for his own interests.

Actually, he invited himself in and his redesign, through the UN and the IPCC, did not stop global warming or climate change, but brought serious global problems. The IPCC identification of CO^2 as the major culprit of environmental damage has:

Allowed an unfounded and unwarranted attack on fossil fuels and exploitation of the false idea we are running out, especially of oil.

Caused governments to promote alternate fuels as if they are the replacement solution when most are not viable alternatives.

Caused governments to provide massive direct or indirect subsidies that distort the value of these alternatives so that accurate cost benefit analysis is essentially impossible.

Caused governments to provide subsides for biofuels that seriously impeded world food production.

Corn based ethanol, causes higher food prices as corn for fuel crowds out corn for food. Corn production is so fossil fuel intensive it does not reduce fossil fuels and it does not even reduce CO_2 emissions. The producer's subsidy and mandate that it be added to gas appear to be politically invulnerable because we hold the first presidential primary in Iowa. Malthusians are phony environmentalists. They don't really care about increased CO_2 emissions from ethanol as long as they can strike a blow at the fossil fuel industry. They want expensive energy to in order to slow economic growth, which they view as cancer eating up the finite resources in their Petri dish.

Caused governments to identify CO_2 as a pollutant and seek its reduction when it is essential to plants and a reduction would put them in jeopardy. When many of our food crops evolved, such as wheat, CO_2 levels were much higher. Therefore these plants are CO_2 starved. The only proven effect of CO_2 is increased plant production. Greenhouses commonly triple the CO_2 level in order to increase plant growth.

Caused many governments to restrict or ban development of most fossil fuel energy sources.

Caused governments to spend billions on climate research to stop climate change when it is impossible.

Caused a diversion of money to climate research better spent on real and identified pollution problems.

Allowed Malthusians bully whole societies into adopting inappropriate policies and ideas.

Caused unnecessary increases in transportation costs that resulted in a higher cost of living that especially impacts the poor and middle class.

Caused extensive and unnecessary fear among people, especially children.

The title of Vaclav Klaus' book *Blue Planet in Green Shackles* succinctly summarizes the broader problem, which he summarizes as follows:

Future dangers will not come from the same source. The ideology will be different. Its essence will nevertheless be identical: the attractive, pathetic, at first sight noble idea that transcends the individual in the name of the common good, and the enormous self-confidence on the side of its proponents about their right to sacrifice the man and his freedom in order to make this idea reality. What I had in mind was environmentalism and its present strongest version, climate alarmism.

There is no scientific justification for any energy or economic policies designed to reduce greenhouse gases or stop warming or climate change. CO_2 from human or natural sources is not causing global warming or climate change. Beliefs that it is are solely the product of the IPCC and their computer models, an agency and approach set up to mislead the world. Yes, as Sir David King said: global warming is worse than terrorism.

I was in the Peace Corps in Brazil, working with sugar cane field hands and their children. They called me Sir David King. Since I think the British Sir David King is trying to march the world into a swamp and I'm trying to turn the march around, I will take my public service "Sir" over his.

We are all environmentalists to a greater or lesser degree. It is an outrage that certain people and groups have usurped this title and implied that only they care about the environment. While this book outlines the role of the Intergovernmental Panel on Climate Change (IPCC) in manipulating climate science, it did this within environmentalism as the new paradigm in the western view of the world.

The message the IPCC pushed suited the Malthusians. It enabled them to hide their activities from the usurped moral high ground of saving the planet with scientific proof. They could isolate those who dated to question the science as anti-environment or paid by the oil companies who were the cause of the major problem of climate change. While this happened, politicians were convinced by the bureaucrats representing their country as members of the IPCC. Politicians were bullied because they didn't understand the science and wanted to appear green.

Primarily due to the actions of the IPCC, nobody tested the scientific theory that human CO_2 caused warming/climate change know and AGW. Rather, Maurice strong set up the organization through the UN to perpetuate the unproven theory. They designed the IPCC mandate so they only looked at human causes of climate change, but the media and the public believe they are looking at natural climate change in total and scientifically. Rules guaranteed the message to the media' they created the illusion that they practiced and accepted only peer-reviewed science.

With a dramatic court defeat in October, 2010, climate alarmists lost a key legal battle. Only once before, in 2007, at the High Court in London, had AGW been challenged in court. The court ruled that Al Gore's film, an Inconvenient Truth, contained nine critical factual errors and could no longer be shown in Schools in England and Wales as a portrayal of fact and was a violation of the British Law against propaganda being used in schools. Among these critical errors was Gore's use of Ice Core data to show that CO^2 and Temperature is correlated. He failed to mention that the temperature started up an average of 800 years earlier than the CO^2. In other words, temperature increases cause CO^2 increases, probably by driving CO^2 out of sea water. In the film Gore said the CO^2 rise caused the temperature rise and said the relationship was complicated. Explaining how CO^2 reached back in time 800 years to cause temperature to rise is indeed complicated and not true as the High Court found.

In 2010, The New Zealand Climate Science Coalition (NZCSC) had demonstrated how certain western national governments arguments for global warming are so easily shredded when employing a long-standing legal tactic available to common law, citizens. In an article by John O'Sullivan "Prosecuting Climate Fraud: The International Dimension he wrote: I explained how such legal triumph could be won. I showed the legal thread that not only linked the five English speaking nations—the US, UK, Canada, Australia, New Zealand, but also provided the key to victory. All such Anglophone nations, while operating their independent legal systems, nonetheless premise themselves on English common law. Under common law our respective governments cannot impose climate regulations on us by regarding similar facts and circumstances differently on different occasions. This principle is known among legal practitioners as stare decisis (judges are obliged to obey the set-up precedents established by prior decisions).

There are ninety legal challenges filed in US courts against the meritless federal climate legislation being brought in via the back door by the Environmental Protection Agency (EPA). It became clear that the EPA sought to impose upon the people 'arbitrary and capricious" governmental, climate-related decisions with little or no scientific justification.

All such challenges are traditionally referred to as mandamus (we mandate) petitions. The New Zealand version of mandamus is known as an 'Article 78' action. New Zealand's National Institute of Water and Atmospheric Research (NIWA) stood accused of repeatedly frustrating NZCSC in its attempts to get government climatologists to explain how they managed to create a warming trend for their nation's climate that is not borne out by the actual temperature record.

NZCSC petitioned the high court of New Zealand to force NIWA (effectively the KIWI government) to validate their national weather service's reconstruction of temperatures—or strike it down. Ostensibly, NZCSC would present evidence in court that NIWA had faked their nation's climate data if they declined to disown it.

Before the matter could be put to the court for a final judgment NIWA's statement of defense gave up the fight. Their attorney's advised the court that NIWA never accepted responsibility for a national temperature record (referred to by them as the NZTR).

Thus by distancing itself from the indefensible NIWA confessed there was never any such thing as an "official" NZ national temperature record, despite there being an official government acronym for it (NZTR). Controverting all previous policy statements, the NZ government now wishes it to be known that the country has never maintained an official record: all such published data was only intended for internal research purposes and not as evidence to prove the country warmed due to human emissions of carbon dioxide.

However, all such data had shamelessly been hyped up via the IPCC as the gold standard of the entire New Zealand temperature history and for decades cited by pro-green advocates as proof of man-made global warming. Along with discredited Australian records, the NZ numbers represented the cornerstone of Australasia/South Pacific warming. Significantly, this region constitutes two of the eight terrestrial Eco zones: we may now infer that at least one quarter of the world's "official" climate record is discredited and an unjustified carbon tax is being extorted."

NZCSC had previously issued a joint press release with the

Climate Conversation Group accusing NIWA of publishing misleading material. Repeated refusal to come clean lead to charges the NIWA had been defensive and obstructive in requests to see New Zealand climate scientist's data.

According to NZCSC, climate scientists cooked the books by using the same alleged trick employed by British and American doom saying scientists. This involves subtly imposing a warming bias during what is known as the homogenization process that occurs when climate data need to be adjusted. Indeed the original Kiwi records show no warming during the 20th century, but after government sponsored climatologists had manipulated the data a warming trend of 1 degree C appeared.

When such data adjustments are made, scientists must keep their working calculations so that other scientists can test the reasonableness of those adjustments. According to an article in *Mathematical Geosciences*, homogenization of climate data needs to be done because non climatic factors made data unrepresentative of the actual climate variation. The great irony is that the justification made for the need to homogenize data is because if it isn't then the conclusions of climatic and hydrological studies are potentially biased. In other words, climate scientists need to add their own spin to the raw temperatures because if they don't then they are less reliable.

However, according to the independent inquiry into Climategate chaired by Lord Oxburgh, it was found that it was the homogenization process itself that became flawed because climatologists were overly guided by subjective bias. Notably, Australian Andrew Bolt, writing for *The Herald Sun*, sagely determined that the Kiwigate scandal was not so much about "hide the decline" but ramp up the rise. Bolt goes on to report, those adjustments were made by New Zealand climate scientist Jim Salinger, a lead author for the IPCC. Salinger was dismissed by the New Zealand government in 2010.

Salinger once worked at Britain's Climate Research Unit (CRU), the institution at the center of the Climategate scandal. Salinger became part of the inner circle of climate scientists whose leaked emails precipitated the original climate controversy in 2009.

In an email to fellow disgraced American climate professor,

Michael "Hockey Stick" Mann, Salinger stated he was "extremely concerned about academic standards" among climate skeptics"!

In 2010, in what seemed like a reprise of Phil Jones debacle at the CRU, the Kiwi government finally owned up the "NIWA" does not hold copies of the original worksheets". Climate audits are slowly unmasking scientific fraud.

Kiwigate follows the pattern of Climategate, the leaked emails from the Climate Research Unit in Britain. First, climate scientists declined to submit their data for independent analysis (called climate audits). Second, when backed into a corner the scientists claimed their adjustments had been "lost". Third, the raw data itself proves no warming trend.

NZCSC explained their frustrations in trying to get to actual truth about what had happened with New Zealand's climate history. The New Zealand government (NIWA) did everything they could to possibly help us, except hand over the adjustments. It turned out that there was actually nothing more could have done- because they never had the adjustments. None of the scientific papers that NIWA cited in their impressive sounding press releases contained the actual adjustments.

After a protracted delay, NIWA was forced to admit it has no record of why and when any adjustments were made to the nation's climate data. Independent auditors have shown that older data was fudged to make the past temperature appear cooler, while modern data was inexplicably increased to portray a warming trend that is not backed up by the actual thermometer numbers.

It is not just one or two nations that the official government climate numbers are awry. Similar such detailed analysis in North America performed by veteran meteorologists Joe D'Aleo and Anthony Watts and published in an SPPI, Surface Temperature Records- Policy Driven Deception, gives cause for concern that we are looking at a worldwide phenomenon. Anthony Watts runs the most popular Climate Skeptics website, WATTS Up With That. This website and others reported on the findings of two independent climate researchers that analyzed data used by the IPCC. The IPCC record showed warming of two degrees per century in Australia that had no scientific explanation. A study by

Willis Eschenbach exposing this arbitrary and capricious adjustment was wholly substantiated by Ken Stewart. What was evident was that NASA GISS, based at Columbia University in New York City, had manipulated a century's worth of Queensland temperature records to reverse a cooling trend in one ground weather station and increase a warming trend in another to skew the overall data set.

The leaked CRU emails address the most recent of the disputed numbers from 2006-2009 and shows how Harry Harris admits government climate data is unusable: "getting seriously fed up with the state of Australian data, so many new stations have been introduced, so many false references, so many changes that aren't documented. I am very sorry to report that the rest of the databases seem to be in nearly as poor a state as Australia was."

These disturbing findings thus call into question both the integrity and methods of government climatologists and have been condemned by UN IPCC Expert Reviewer, Dr. Vincent Gray. Gray has been a UN IPCC Expert Reviewer for four UN IPCC reports: 1991, 1995, 2001 and 2007.

Dr. Gray confirms that the raw temperatures, free of the chicanery of governmental "homogenization" exhibit no such warming bias.

In addition we see that Dr. John Christy, University of Alabama-Huntsville published two detailed studies that demolished the American "homogenized" records similarly derived from the same NASA-GISS data sets.

But apart from fiddling the temperatures already in their possession, climate fraudsters sought to manufacture a warming bias in the future by causing the "disappearance" of 806 inconvenient cooler weather stations around the world. All 806 weather stations were dropped from the total of 6,000 temperature stations in a single year with no explanation from the Global Historical Climatology Network (GHCN), the government organization that maintains this data and which is used by the UN and worldwide governments.

When we forego the homogenized government numbers and go instead to the primary source of accurate thermometer readings, we get a different picture with no apparent man made warming.

Two such accurate raw data sets are the world's oldest and most reliable; they are Britain's Central England Temperature record (CET) and Central European set from Klementinium at Prague in the Czech Republic. Dr. Jan Zeeman has written a paper that shows they is no human signal in the European record since the 1790s in central Europe is a mere quarter of one degree (.0.265C per century).

The Prague raw temperatures correlate extremely well with the Central England Temperature record (CET) that has been running continuously for 351 years (1659-2009).

What we see is neither in central Europe nor in England has there been any signal of man-made warming in the recorded instrumental history. As these datasets are considered the best proxies for Northern Hemisphere temperatures and since global temperature trends follow a similar pattern to Northern Hemisphere temps then the same conclusion on AGW can be inferred globally.

Rather than publish the facts, include the raw data in the analysis of AGW, the IPCC has instead chosen to misrepresent to international policy makers a pattern of inexplicable warming by reference to the homogenized numbers—the skewing of which cannot be accounted for from the raw temperature data.

Thus for two decades policy makers were presented with a consistently false picture indicating a warming trend that only existed in the "laboratories" or "fudge factories" of climatologists.

Thus we have identified the true source of man-made global warming. It is the clandestine number falsification published in the IPCC summary reports delivered to national governments and world media.

Suspicions grew that British and EU governments were fudging climate data to fulfill a predetermined goal. Their climatologists were being paid to create the illusion of human induced climate change. Several Freedom of Information requests (FOIA) were filed by independent analysts over the years, most famously by Canadian statistician, Stephan McIntyre.

McIntyre, who was retired, had developed a process he called a climate audit. He merely audited or checked the math of various assertions about APG. He challenged the NASA-GISS data set by analyzing the adjustments that they had made to temperatures. He found that there were simple errors in the math they said they were using to homogenize the raw data. One of these adjustments was to the US temperature in 1934. NASA (Hansen) admitted their math error and the corrected temperature for 1934 remains the highest average temperature year in the United States. This fact means that no one in the United States under the age of 80 has experienced net warming in the US.

Professor Jones of the Climate Research Unit (CRU), University of East Anglia, home of Climategate and the world's leading center for climate data homogenization, instructed his colleagues to destroy all such data and not submit it to McIntyre's lawful FOIA requests. Jones was targeted for criminal investigation due to his unequivocal admissions of misconduct in the leaked Climategate emails. The subsequent official investigations by the UK Information Commissioners' Office (ICO) substantiated the claim that potentially incriminating calculations (metadata) formulated by government researchers in the homogenization process had been destroyed—a criminal act.

Leaked emails written by Jones proved he threatened to destroy his data rather than allow McIntyre to see it. When the ICO investigated they discovered Jones had, indeed destroyed the data. Apologists for the crime assert that Jones did not destroy original raw temperature records.

This may be true: however, Jones did destroy his adjustments that would have been key evidence as to his intentions to commit climate fraud. Legal analysts argue the destroyed evidence would likely have proved Jones et. al. acted with fraudulent intent. Indeed, statistical forensic experts affirm that if they had been allowed to have examined the data before Jones destroyed it, than any unwarranted adjustments could be readily identified as being caused by faulty system programs or on a one by one basis consciously manipulated with the intention to fraudulently deceive. The ICO determined that the short stature of limitations, six months, had already expired.

Thus climate scientists in different countries at the opposite side of the world are facing extraordinarily similar data fraud allegations. With so many climatologists having "lost" their calculations, no one can now replicate their methods or have any confidence in the claims that mankind has unnaturally warmed the planet.

The trend is now undeniable: the man-made global warming in the 20th century comes not from the raw data of thermometer readings, but through the man-made tampering perpetuated inside leading climate institutions like the Climate Research Center. Judging by the raw temperature data alone and taking into account the fraudulent adjustments to the raw data, we must conclude that there is no evidence of unusual planetary surface warming in the last century.

In Nov. 2010, Scientific American announced the results of their recent opinion poll of it scientifically literate readership. A total of 83% of 5190 respondents think the UN's IPCC is "a corrupt organization, prone to group-think, with a political agenda.

Chapter 26

THE PRESIDENT'S COUNCIL ON SUSTAINABLE DEVELOPMENT AND THE NATIONAL PERFORMANCE REVIEW

President Clinton signed Executive Order (EO) 12852 on June 29, 1993. The EO established the President's Council on Sustainable Development and appointed Al Gore to conduct a National Performance Review. The council seated 25 members including most of his cabinet secretaries, representatives from strong AGW proponents such as The World Resources Institute, Natural Resources Defense Council, and the Women's Environment and Development Organization. Working in tandem with Gore's National Performance Review, an overhaul of the Departments of Interior and Agriculture were conducted to implement what Gore called the "Ecosystem Management Policy". This new protocol was coordinated with existing legislation, such as the Endangered Species Act and the Clean Water Act, authorizing the Federal Government supreme power to regulate land use in rural America. Additionally, the Ecosystem Management Policy provided state and local governments with financial incentives to aggressively regulate land use, and it implemented "stakeholders".

This clause instituted civilian councils that work with the government in public-private partnerships. These stakeholder councils are typically groups like the Nature Conservancy, Sierra Club, Water Guardians, World Wildlife Fund or Greenpeace, allowing them a seat at the table without ever having to be elected to public office. The duty of the stakeholders to create public pressure to enforce and create local, state, and federal environmental law: raise private money; and garner further public funding—all with the ultimate goal of declaring large swaths of land off-limits to development. Stakeholder councils have been known to use shakedown tactics, sue their partners the government, abuse eminent domain, conjure claims of endangered species, and procure phony scientific studies to achieve their desires.

Stakeholders, and their like-minded co-conspirators working within all levels of government, commonly refer to a manual created by Gore's Ecosystem Management Policy: *The Growing Smart Legislative Guidebook: Model Statutes for Planning and the Management of Change. Growing Smart* provides detailed instructions on how to create legislation, implement tax policies, and use existing laws and regulations to squeeze private landowners. Tactics from the guidebook have provided the template to assist in the creation of needless open-space preserves—all with the goal of removing mass acreage from future private development.

Chapter 27

SUSTAINABLE DEVELOPMENT

The President's Council on Sustainable Development in 1995 announced a report with:

We believe, economic growth, environmental protection and social equity are linked. We need to develop integrated policies to achieve these national goals.

The United States should have policies and programs that contribute to stabilizing global human population.

Remember the policies for stabilizing populations that President Obama's science advisor wrote about. Will he recommend forced abortions and adding sterilant to drinking water as our Foreign Policy? Remember, Erlich said, Giving cheap abundant energy to people was like giving machine guns to idiot children?

Now the state plans to take your cheap and abundant energy away from you. Who did you think the idiot children were, pygmies in Borneo? Let me now apologize to all the pygmies in Borneo, who I think are far too wise to fall for the Climate Change Hoax.

In 2008, the California Energy commission proposed revisions to the 230 page California Building Code. These proposed building codes mandated that remote controlled home thermostats be installed in all new or remodeled homes or in existing homes in which the furnace or air conditioner was being replaced.

This remote controlled home thermostat is called a Programmable Communication Thermostat (PCT). Embedded within the PCT is an FM radio receiver, which would allow energy authorities to control home temperatures. Customers would be denied the ability to override settings. The PCTs would help avoid the "rolling blackouts" often experienced during periods of peak power demand in California.

Because we have a law mandating a third of energy in California comes from renewables, California leads the nation with 15% of our energy coming from wind and solar. When the wind doesn't blow and the sun doesn't shine, our base load is generated from natural gas plants which supply about 64% of California electricity. Because of the goal of increasing use of wind and solar, the gas plants are not operated at their full capacity. This underutilization makes them unprofitable. Thus utilities are shutting them down. As we lose our base load, we will experience more rolling blackouts. So the central planners have their solution, PCT will take your cheap and abundant energy out of the hands of idiot children, i.e. you.

With PCT technology, the thermostat czar would regulate the settings of every air conditioned by electricity and natural gas heated home in the state.

You would not make the decision of where to set the thermostat since you are an idiot child. There would be public service announcements explaining your civic duty to wear warm clothes inside your home, because we have to conserve energy during this Climate Change crisis, which the Malthusians wearing their green masks engineered in order to take cheap and abundant energy from idiot children.

The PCT is part of an emerging technology for direct load control. The technology allows the utility company to charge customers for electricity based on the time of day usage and provides the company to turn down, or even interrupt the power flowing into a customer's home.

There was a firestorm of protest and the commission backed off their plans. However, a staff member said they would eventually find a way to get PCT's in every home, because it was part of a much larger plan.

This plan is called smart technology and with smart meters and smart PCT the intention is to eventually ensure that all major appliances, including washers, driers, dishwashers, refrigerators, water heaters and lighting can be governed remotely.

Smart technology was introduced in Federal Regulations is 2005:

"Not later than 18 months after the date of implementation of this paragraph, each electric utility shall offer each of its customer classes a time based rate schedule under which the rate charge by the electric utility varies during different time periods."

The smart meter replaces the old spinning wheel electric meter, which was read monthly by a utility company worker. Utility companies can read smart meters to know how much power you use minute by minute.

In California households were startled to hear that they were all going to get smart meters to replace their old analog meters. There was a process to let you opt out, at least at first, but the default was you were going to get a smart reader. There was a bit of an uproar and confusion, because no local or state governments had taken any votes on this expensive program.

What had happened is that Al Gore had made an investment in a firm that made smart meters. There is a private demand for smart readers. They allow businesses to monitor and control their energy use. This is a relatively small market compared with mandating all households in California get a smart meter.

Gore used his connections in Washington to get smart meters incorporated into the stimulus funding bill. In a program which was intended to increase employment by stimulating the economy, it was a bit odd those smart meters, which replace meters readers, and thus cost jobs, was included.

Al Gore left the White House with fewer than two million dollars; his net worth now is between 100 and 500 million dollars. By combining his political connections with a green mask he has made a remarkable fortune in the world of green crony capitalism.

In 2009, General Electric (GE) received an award from the US Department of Energy and the EPA, as the Federal Government's Energy Star Partner of the Year for their advances in "smart" or Management enabled devices designed to modulate power consumption based on usage time and rates.

"Management Enabled" is trade speak for appliances that can be managed remotely. This is marketed as if the consumer is the only one that can manage appliances remotely.

Darrel Bracken a regional president of Whirlpool, announce that by 2015, their company will "make all the electronically controlled appliances it produces—everywhere in the world—capable of receiving and responding to signals from smart grids.

Since California is driving its wind and solar to provide 30% of electricity generated, and part of this plan is to not use natural gas plant at their optimum levels which makes them unprofitable and results in closing the plants that provide 64% of our base load, rolling black outs will result unless we implement a smart grid. The people of California will be forced to use less electricity by force, either by pricing or direct control.

Under certain predetermined peak period, the utility interrupts or cycles the appliances to achieve its system goal of reducing peak usage. In most cases, the Malthusians hope the customer will not notice the adverse consequences. After all the idiot children probably won't even notice their dishes didn't get clean and their clothes didn't get dry.

California in 2015 is in the midst of a significant drought and regulations about when and how you can water your lawn are growing like weeds. A company named Blossom offers a new smart sprinkler controller to save on water bills. How long before California Water districts, which for years have denied water meters to new customers; require them to get a smart meter that can be controlled by the Water district.

The Smart Grid was created by the Energy Independence Security Act of 2007. A memo from the Congressional Research Service said:

The term SmartGrid refers to a distribution system that allows for flow of information from a customer's home in two directions: both inside the home to thermostats and appliances and other devices, and back to the utility. It is expected that grid reliability will increase as additional information from the distribution system is available to utility operators. This will allow for better planning and operations during peak demand. For example, new technologies such as a Programmable Communicating Thermostat (PCT) could be connected with a customer's meter through a Home Area Network allowing the utility to change the settings on the thermostat based on load or other factors.

The Smart Grid, the Smart Home, the Internet of Things, like all technology can be used for good or ill. The mantra of the Malthusians and AGW true believers is "A crisis is a terrible thing to waste". Whatever the crisis, this technology is the Trojan horse which will let them enforce whatever limits they wish on the idiot children who are squandering the finite resources in their imaginary Petri Dish.

Why should anyone need more than five minutes to take a hot shower? Why should anyone have a green lawn, an offensive waste of water and fertilizer? Why should you be able to wash and dry your clothes, wash your dishes, air condition and light your house whenever you please? We must enforce mandatory regulations to conserve because idiot children won't make the necessary sacrifices. Obviously people drive too much. A smart controller on your car could ration your mileage to 25 miles a day. Anyone who needs to travel farther should take public transportation.

California Utility giant PG&E has a program called SmartTag. This device would not replace the thermostat, but is located outdoors, next to the customer's air conditioner. PG&G says:

When activated, your AC will do what it would normally do for about 15 minutes of every half hour-make cool air. Your system fan will circulate already cool air and your AC will make cool air again when the next 15 minute cycle begins. You'll not even notice when this happens. In fact, in a survey of sample participants, most didn't even notice a change.

This is how the Trojan horse of smart devices will be introduced, "you won't even notice the change". Of course the SmartAC can be controlled by PG&E who can enforce mandatory conservation whenever they please. Why should PG&E provide cheap and abundant energy to idiot children, when it is much more efficient to cut their consumption as required?

Contained in the 2009 America Clean Energy and Security Act are federally mandated energy-efficient building regulations, which supersede all local and state codes, "to provide necessary enforcement of a national energy efficiency building code".

In California, people from cold states sometimes say the wind blows right through your houses. Should the same energy building code be enforced from Alaska to Florida, from Hawaii to Maine by the Federal Government? The Federal Government has the implied legislative authority to issue federal regulations that will require that every time a home is built, remodeled and, or prepared to be sold, you will be required to have a smart thermostat, smart electric meter, and all your smart star appliances will have to be tied to a Home Area Network which can be recorded and controlled by bureaucrats in Washington, DC.

We see weather forecasts on TV daily. How many would make an important decisions based on a forecast 7 days away. When it comes to "Climate Change", the $360 billion dollar a year climate establishment is telling us that civilization must be reorganized from top to bottom based on failed models purporting to make predictions decades and centuries in advance. Flawed predictions aside, a great deal of evidence suggests accuracy or truth was never the intent. Alarmists have often written that they had to choose between being effective and being honest. Effective to them means generating fear to seize more money and power. Some alarmists have responded to the implosion of their theories with calls for censorship and some papers have said they will not publish letters critical of AGW. Some have even called for Nuremburg type Climate Trials leading to imprisonment, re-education or even execution of "climate deniers".

They say we must transition to the Valhalla of renewables from fossil fuels now. California's Governor Jerry Brown, is demanding a goal of increasing California 15% renewables to 50%. Meanwhile, he is stealing millions raised from CO_2 cap and trade program to fund an $86 billion dollar bullet train. Cap and trade programs are carbon taxes and big ones that politicians can't wait to get their hands on. However, you can't fool all the people all the time. Climate Change alarmism is slowly going the way of the Y2K false alarm. Historians will have a great time explaining why the climate con worked so well, for so long.

Chapter 28

THE CLIMATE BOTTOM LINE

The *New York Times* ran an article February 1, 2015, "The Climate Bottom Line—Flooded properties. Lower Crop Yields. Stalled Trains. Climate change is predicted to have big economic costs, and an influential group wants business to be ready." The point of this book has been to document that there is no evidence that flooding is increasing, that crop yields are falling and that trains are stalling due to climate change. Please note that the predicted Biblical catastrophe is now Noah's flood. Notice also that this prediction has no timeline. The climate alarmists have learned how pathetic they look when they give a date, such as by 2020 there will be increasing flooding and it fails as all their predictions have failed since 1988 when they gave a timeline. Nevertheless, I would be quite content to let this nonsense join the dustbin of history except for the fact that Malthusians have been wrong for 216 years and there are more Malthusians, wearing the mask of climate alarmists, than ever.

Consider Greg Page, the executive Chairman of Cargill, the food conglomerate. He says that over the next 50 years, if nothing is done, crop yields in many states will most likely fall, the costs of cooling chicken farms will rise and floods will more frequently swamp the railroads that transport food in the United States.

Mr. Page is a great man and as Chairman of Cargill, is responsible for a significant proportion of the food supply of the globe. However, we all due respect, Mr. Chairman, what the hell are you talking about? The day Mr. Page was interviewed for *The New York Times* article it was 8 degrees in Minneapolis and snowing. Let's do a mental experiment. Instantaneously double the CO_2 and increase the temperature by 2 degrees F. Now the temperature is 10 degrees F and it is still snowing.

Chicken farms costs of heating will fall and costs of cooling will rise and the net difference will likely be beneficial. Due to the increase of CO_2, crop yields will increase and flooding will remain the same. Robert Mendelsohn and many other economists have concluded that a rise of 2 degrees F will result in benefits exceeding costs.

Compare this with the Kyoto Protocol. Assume that every nation signed the treaty and accomplished the draconian cuts in CO_2 emissions. According to the treaty, in 50 years the temperature would be affected by .17 degrees F. In other words, at the cost of trillions of dollars in economic growth we would reduce temperatures by a fraction of 1 degree. The costs clearly exceed the benefit. Mr. Chairman it could certainly get hotter, it has been hotter than it is today 97% of the time for the last 10,000 years. However, cutting CO_2 emissions is the most expensive way to adapt to global warming. Whether you are going to heat or cool chicken farms, cheap and abundant energy is the cost driver. Mr. Chairman, that is the climate bottom line.

www.ingramcontent.com/pod-product-compliance
Lightning Source LLC
Chambersburg PA
CBHW060602200326
41521CB00007B/641